# SNOWSTORMS ALONG THE NORTHEASTERN COAST OF THE UNITED STATES: 1955 to 1985

# METEOROLOGICAL MONOGRAPHS

Inquiries about the above publications should be sent to:

*THE AMERICAN METEOROLOGICAL SOCIETY*
*45 Beacon St., Boston, Mass. 02108*

## METEOROLOGICAL MONOGRAPHS

Volume 22     April 1990     Number 44

# SNOWSTORMS ALONG THE NORTHEASTERN COAST OF THE UNITED STATES: 1955 to 1985

## PAUL J. KOCIN AND LOUIS W. UCCELLINI

*Goddard Laboratory for Atmospheres*
*NASA/Goddard Space Flight Center*
*Greenbelt, Maryland*

American Meteorological Society

**Library of Congress Cataloging-in-Publication Data**

Kocin, Paul J.
  Snowstorms along the northeastern coast of the
United States, 1955 to 1985.

  (Meteorological monograph; no. 44)
  Includes bibliographical references.
  1. Blizzards—Northeastern States.  2. Blizzards
—Atlantic Coast (U.S.).  I. Uccellini, L. W.
II. Title.  III. Series: Meteorological monographs
(American Meteorological Society); no. 44.
QC929.S7K59   1990     551.57′843′0974     90-376
ISBN 0-933876-90-4

ISBN 0-933876-90-4

Typeset and printed in the United States of America by Lancaster Press, Lancaster, Pennsylvania.

Published by the American Meteorological Society, 45 Beacon Street, Boston, MA 02176.

Richard E. Haligren, Executive Director
Kenneth C. Spengler, Executive Director Emeritus
Evelyn Mazur, Assistant Executive Director
Arlyn S. Powell, Jr., Publications Manager
Jon Feld, Publications Production Manager

Editorial support provided by Linda Esche, Leslie Keros, and Susan McClung.

## Dedication

*Paul J. Kocin dedicates this monograph to his wife Marla, his sons Matthew and Joshua, his parents, Eugene and Irene Kocin, and to his sister, Barbara Wallace.*

*Louis W. Uccellini dedicates this monograph to his parents Louis and Margaret Uccellini, his wife Susan, his children Anthony, Francesca and Dominic and his friend Frank DeLaurentis who shared many a snowstorm on Long Island and is a retired snowplow operator living near Troy, NY*

# Table of Contents

# ACKNOWLEDGMENTS

Many individuals have contributed directly and indirectly to the preparation of this monograph. We would like to thank Kelly Pecnick, of the Severe Storms Branch of the Goddard Laboratory for Atmospheres at NASA/Goddard Space Flight Center, for enduring the monotony of typing countless revisions of the manuscript. Keith Brill and Robert Aune provided computer programming support and data tape reduction techniques. Lafayette Long and William Skillman provided assistance in obtaining photographic products and in acquiring data, while Deena Acton of the Library Services Branch at NASA/GSFC helped retrieve microfilmed records of newspaper accounts of the storms. Members of the Presentations Graphics Section provided technical assistance in preparing artwork and photography.

William Blumen, editor of the AMS *Meteorological Monographs,* Chester Newton, and two anonymous reviewers provided constructive reviews which helped to improve the manuscript. We are especially grateful to Alan Weiss, who provided the most thorough review of any individual involved. Informal reviews by Daniel Keyser, John Zack, Lance Bosart, Stephen Colucci, and Thomas Matejka were also extremely important in shaping the final form of the monograph. David Ludlum, Chester Newton, John Zack, Lance Bosart, Daniel Keyser, Kenneth Spengler, Joanne Simpson, Robert Atlas, and Alan Weiss are singled out for their inspiration and sincere desire to see this project completed.

Other individuals who contributed to this project include Katherine Gratke in Madison, Wisconsin; and Andrew Horvitz; William Poust; and Bruce Needham of the Satellite Data Services Division of the National Environmental Satellite, Data, and Information Service/National Climatic Data Center in Camp Springs, Maryland, who are recognized for their efforts in locating satellite imagery from TIROS, ESSA, ATS, and SMS-GOES satellites. Special notes of appreciation are also extended to Joseph Moyer, the State Climatologist for Maryland and Delaware, to Vincent Cinquemani of the National Climatic Data Center in Asheville, North Carolina, who assisted in our search for snowfall records throughout the northeastern United States, and to the staff members of the NOAA library in Rockville, Maryland, who were instrumental in obtaining necessary references. Last, but not least, we thank James Dodge of NASA Headquarters for providing financial support throughout this project.

# Prologue

Saturday, 8 February 1969, was not the bitterly cold day that often serves as a harbinger of a severe winter snowstorm on New York's Long Island. Instead, it was generally sunny and cool, about 40°F (5°C). The weather forecast that morning, though, provided a little bit of encouragement to a 13-year-old snow fanatic: "Tomorrow will be cloudy with rain OR SNOW likely."

I[1] knew from experience that the probability of a big snowstorm on Long Island was not very great because of its close proximity to the relatively warm Atlantic Ocean. I had already felt the crushing disappointment when potentially major snowstorms became rainstorms, and sure days off from school became days like all the rest. My pessimism was reinforced by the local weather reports on Saturday evening, indicating southeasterly winds, a bad sign for impending snowstorms. As I went to bed that night, I prepared myself for the usual disappointment, since outdoor temperatures held well above freezing and the wind continued to blow from the ocean.

When I awakened the following morning, a strong wind was whistling through the air-conditioning unit that extended outside my room. This sound was not enough to arouse me until I suddenly realized that the glow emanating from the translucent shutters covering my window was unusually bright for early morning. I shot out of bed and flung open the shutters to discover that the world outside was white. There wasn't very much accumulation on the ground yet, maybe an inch, but the snow was falling fast and flying horizontally from the northeast. A busy day lay ahead of me!

While my family still slept, I ran from window to window to see how the storm was changing the environment outside. I turned the radio on, only to become annoyed at the weather forecasters' insistences that the snow would soon end. After a while, confidence in the original forecasts eroded as the storm grew more and more intense.

By late morning, my family awoke in astonishment as they gazed out the window, wondering what to do now that they knew the day would be spent inside. The snow had picked up in intensity, falling at perhaps 1 or 2 inches (2.5 to 5 cm) per hour. Savoring every moment, I felt victorious when the weather forecasters finally succumbed to the storm and admitted that they didn't know how much more would fall or when the storm would end. Then, the inevitable comparisons with earlier blizzards began to pour forth from the radio and TV.

The day wore on and snow continued to fall heavily. The wind increased, making it difficult to determine how much of the swirling snow outside was due to falling snow versus snow blowing off the roof or from the ground. Drifting snow eventually began to cover several windows, making it more difficult to observe what was obviously a great storm. As the drifts grew, it became impossible to open the front door, and by evening, the cars parked on the street appeared as white swells in the sea of snow that

[1] Paul J. Kocin

covered the street. Tremendous gusts of wind whistled through the house and actually seemed to shake the foundation.

The snow finally ended in the middle of the night. About 18 hours had passed since the time the early forecasts indicated that the storm would cease. By morning, the only sound heard was that of an overworked snowplow. Radio reports now documented the extent of the storm. The entire New York City metropolitan area was paralyzed. As several hundred cars became stranded, scores of motorists spent a harrowing night on the Tappan Zee Bridge north of New York City. Many other motorists met similar fates on the myriad of roadways throughout the metropolitan area that were littered with stranded cars. Thousands more were stranded at airports that remained closed anywhere from 1 to 2 days following the storm. Food was airlifted by helicopter to Kennedy Airport to feed the weary travelers. Schools closed on Monday and many didn't reopen until Thursday, Friday, or even the following Monday. Snow-clearing equipment was slow to be deployed in parts of the New York City area and couldn't keep up with the storm. Some city streets remained unplowed many days after the storm, creating a public furor. Dozens of deaths and several hundred injuries were attributed to the storm while millions of dollars were lost due to delayed or lost business.

This particular snowstorm, and other similar storms during the late 1950s and 1960s, provided the motivation for the monograph that follows. Questions concerning how these storms develop, what weather patterns provide clues that foretell such events, and what factors delineate snow/no snow situations, have challenged forecasters and researchers alike. This study is our attempt at describing a phenomenon that has stirred our curiosity, and provided countless hours of speculation, entertainment, and, for those numerous false alarms, profound disappointment. The disappointments (i.e., those storms that changed to rain or veered harmlessly out to sea) will not be highlighted here. The greatest snowstorms to affect the northeastern coast of the United States during the 1950s, 1960s, 1970s, and 1980s will be explored in the following chapters.

# CHAPTER 1

# Introduction

The aesthetic appeal of snowstorms is shared by many, who, overlooking the discomforts and hardships such storms may yield, find the transformation of familiar surroundings into spectacular landscapes of white and gray an exhilarating experience. The strong emotions brought forth by these great storms of snow and wind are revealed in the following excerpts:

> There is nothing quite like the promise and anticipation of a good snowstorm. It makes Christmas pale, birthdays insignificant, and is rivalled only by Election Day for the joy of anxious hope.[1]

> The sun that brief December day
> Rose cheerless over hills of gray,
> And, darkly circled, gave at noon
> A sadder light than waning moon.
> Slow tracing down the thickening sky
> Its mute and ominous prophecy,
> A portent seeming less than threat,
> It sank from sight before it set.
> A chill no coat, however stout,
> Of homespun stuff could quite shut out,
> A hard, dull bitterness of cold,
> That checked, mid-vein, the circling race
> Of life-blood in the sharpened face,
> The coming of the snow-storm told.
> The wind blew east; we heard the roar
> Of Ocean on his wintry shore,
> And felt the strong pulse throbbing there
> Beat with low rhythm our inland air.

> Meanwhile we did our nightly chores,
> Brought in the wood from out of doors,
> Littered the stalls, and from the mows
> Raked down the herd's-grass for the cows:
> Heard the horse whinnying for his corn;
> And, sharply clashing horn on horn,
> Impatient down the stanchion rows
> The cattle shake their walnut bows;
> While, peering from his early perch
> Upon the scaffold's pile of birch,

---

[1] From an article entitled "Meteorological Ambitions" from *News from Bennett Noble.* Source unknown.

The cock his crested helmet bent
And down his querulous challenge sent.

Unwarmed by any sunset light
The gray day darkened into night,
A night made hoary with the swarm
And whirl-dance of the blinding storm,
As zigzag, wavering to and fro,
Crossed and recrossed the winged snow;
And ere the early bedtime came
The white draft piled the window-frame,
And through the glass the clothes-line posts
Looked in like tall and sheeted ghosts.

So all night long the storm roared on:
The morning broke without a sun;
In tiny spherule traced with lines
Of Nature's geometric signs,
In starry flake, and pellicle
All day the hoary meteor fell;
And, when the second morning shone,
We looked upon a world unknown,
On nothing we could call our own.
Around the glistening wonder bent
The blue walls of the firmament,
No cloud above, no earth below,—
A universe of sky and snow!
The old familiar sights of ours
Took marvellous shapes; strange domes and towers
Rose up where sty or corn-crib stood,
Or garden-wall, or belt of wood;
A smooth white mound the brush-pile showed,
A fenceless drift what once was road.[2]

Episodes of heavy snowfall, however, are also cause for concerns that are larger in scope than the mere discomfort and inconvenience of shoveling the driveway or walks. The combined effects of heavy snow, high winds, and cold temperatures can be particularly debilitating, especially in a heavily populated and highly industrialized environment. Such is the case along the Northeast coast of the United States, a region which will be defined in this monograph as extending from Virginia to Maine and includes the densely populated metropolitan centers of Washington, D.C., Baltimore, Maryland, Philadelphia, Pennsylvania, New York, New York, Boston, Massachusetts, and dozens of smaller urban areas. Here, heavy snowfall associated with storms known as "Nor'easters" may maroon millions of people at home, work, or in transit, severely disrupt human services and commerce, and endanger the lives of those who venture outdoors.

Snowstorms along the northeastern coast of the United States result from a complex interaction of physical mechanisms that acts to organize a widespread region of ascent, entrain large amounts of water vapor necessary to produce heavy amounts of

---

[2] From "Snowbound" by John Greenleaf Whittier, *The Family Book of Verse,* Louis Gannett, Ed., Harper and Row, New York, 1961, pp. 302–303.

precipitation, and supply a source of cold air that enables precipitation to reach the ground as snow. These physical mechanisms are associated with cyclonic and anti-cyclonic weather systems that traverse the middle latitudes and interact with the unique topography of the eastern United States.

Topographical factors that contribute to these weather systems include land–ocean temperature and frictional contrasts, the influences of the Atlantic Ocean, Gulf of Mexico, and Great Lakes on air mass modification, the position of the Gulf Stream, and the effects of the Appalachian Mountains on the low-level temperature and wind fields. This combination of factors favors the northeastern United States and its offshore waters as an important site for cyclogenesis (Petterssen 1956; Klein and Winston 1958; Reitan 1974; Colucci 1976; Sanders and Gyakum 1980; Hayden 1981; Roebber 1984) that can occasionally result in heavy snowfall.

The ability of weather forecasters to recognize and account for the many elements that influence the development and evolution of major snowstorms is crucial for the accurate prediction of heavy snowfall in the northeastern United States. Before the advent of numerical weather modeling techniques, attempts to forecast such storms had yielded mixed results, at best. Some of the worst snowstorms in New York City's history, most notably the March 1888 "Blizzard of '88" (Kocin 1983), the city's record snow of December 1947[3], and the "Lindsay Storm" of February 1969,[4] described in the Prologue, were severely underforecast. Thus, the public's continuing lack of confidence in forecasters' abilities to predict such storms is, to some degree, justified by past experience.

The increasing use and sophistication of numerical weather prediction models has improved the forecasts of the timing and location of cyclogenesis, precipitation amounts, and the location of the rain–snow line, all important considerations in potential heavy snow situations. Operational and research numerical weather forecasts have performed well in many recent heavy snow situations, as demonstrated by simulations of the intense New England snowstorm of February 1978 (Brown and Olson 1978), the spring blizzard of April 1982 (Kaplan et al. 1982), and a convective snow event in March 1984 (Kocin et al. 1985). The models have failed on other occasions, however, most notably the February 1979 "Presidents' Day Storm" (Bosart 1981; Uccellini et al. 1984, 1985, 1987; Atlas 1987). An example of a numerical weather forecast that can be described as both a success *and* a failure is the operational numerical model forecast of the urban snowstorm of 11–12 February 1983. A major snowfall was predicted for the Middle Atlantic states (a success), but a small error in the model forecast of the northward extent of heaviest precipitation amounts likely resulted in the underprediction of excessive snowfall in New York City and southern New England (a failure). Therefore, even a relatively good numerical model forecast may contain small errors that can have a profound impact on the accuracy of weather forecasts that are released to the general population.

The inconsistent quality of the forecasts of these storms suggests that certain physical features and dynamical processes, whose influences vary widely from case to case, are better understood or more accurately simulated than others. In view of the

---

[3,4] Forecasts for these storms were examined from local newspaper reports.

obvious need for an improved understanding of these storm systems, this monograph provides an analysis of the horizontal and vertical structures and evolution of 20 of the most crippling snowstorms to affect the heavily populated Northeast region over a 30-year period from 1955 through 1985. These events are examined from historical, climatological, and dynamical perspectives.

It is not the intent of this monograph to introduce detailed dynamical interpretations of cyclone events, as can be found in specialized diagnostic and modeling studies of individual cases. The Genesis of Atlantic Lows Experiment [GALE (Dirks et al. 1988)] and the Canadian Atlantic Storms Program [CASP (Stewart et al. 1987)], conducted in January through March 1986, and the Experiment on Rapidly Intensifying Cyclones over the Atlantic [ERICA (Hadlock and Kreitzberg 1988)], conducted during the winter of 1988/1989, should serve as excellent sources of specialized studies of cyclogenetic episodes that attempt to address the relative importance of the various processes that influence cyclone development along the East Coast of the United States.

Brandes and Spar (1971) attempted to define the necessary conditions for heavy snowfall along the East Coast using a composite approach. They found that compositing failed to resolve antecedent patterns 12–24 hours prior to heavy snowfalls. Apparently, the synoptic- and mesoscale features in the geopotential height and wind fields that are important factors in the evolution of snowstorms were eliminated by combining many storm systems characterized by significant case-to-case variability. Therefore, the approach used in this monograph is based on sequences of conventional weather charts and detailed descriptions of each individual storm. This method enables us to identify patterns in the surface, lower-tropospheric, and upper-tropospheric fields that precede and accompany the development of heavy snowfall and to document the case-to-case variability that frustrates efforts to specify required conditions or standard scenarios that are applicable to all storms.

The interactions of the varied mechanisms and processes that produce these storms span the spectrum of time and length scales from the nucleation of cloud particles to planetary wave dynamics. The temporal and spatial resolution of the operational weather analyses from which this study is derived limit us to describing only those features which operate over periods of several hours to several days and lengths of $10^2$ to $10^4$ km [the meso-, synoptic, and planetary scales; see Orlanski (1975)]. Details on the physical principles in snow formation can be found in Byers (1965) and other texts on cloud physics, and will not be addressed here.

We hope that this work will provide a foundation for researchers and students interested in investigating the processes that interact to produce major winter storms and serve as an easily referenced guide for forecasters concerned with predicting heavy snowfalls along the northeastern coast of the United States. In the following pages, historical and climatological overviews of heavy snow events in the Northeast will be discussed in Chapters 2 and 3. A description of the meteorological characteristics of 20 major storms will follow, derived from conventional analyses of sea-level pressure, geopotential height, wind, temperature, and other parameters, in Chapters 4 and 5. The patterns presented by these fields reveal many of the dynamical and physical processes that contribute to the storms, which shall also be discussed. Chapter 6 provides a brief summary of observations and an overview of the physical processes that contribute to these storms, as well as a discussion of remaining issues. Schematic diagrams illustrating key dynamical patterns and physical processes are included in Chapters 4, 5, and 6, adding a visual perspective to the written descriptions of how the atmosphere evolves prior to and during periods of heavy snow in the northeastern urban corridor. Finally, a series of weather map analyses for each of the 20 storms is presented in Chapter 7, concluding in Chapter 8 with analyses of three storms which occurred during the 1986/1987 winter season.

# CHAPTER 2

# Historical Overview: A Brief Review of Major Snowstorms of the Eighteenth, Nineteenth, and Twentieth Centuries

Through the centuries, severe winter storms have become part of the historical folklore of many regions. A comprehensive accounting of major winter storms in the Northeast over the past three centuries has been assembled by David Ludlum (1966, 1968, 1976, 1982, 1983).[5] His unique studies document the distribution, depth, and duration of great snowfalls, along with their wind velocity, degree of cold, and effects on the general population. Nearly 300 years of such information provides important historical and climatological insight into the limits of severity these storms are capable of attaining. A sampling of the region's legendary storms is briefly described below.

## 2.1 EIGHTEENTH CENTURY

*27 February–7 March 1717.* "The Great Snow of 1717." A series of four snowstorms, two relatively minor and two major, left depths in excess of 90 to 120 cm across most of southern New England, with drifts exceeding 750 cm. The effects of these storms were so impressive that local historical accounts still singled out this storm period as "The Great Snow" more than 100 years following the event.

*24 March 1765.* This storm affected the area from Pennsylvania to Massachusetts. From Philadelphia came this report: "On Sunday night last there came on here a very severe snowstorm, the wind blowing very high, which continued all the next day, when it is believed there fell the greatest quantity of snow that has been known for many

---

[5] Historical discussions of storms also appear in selected issues of *Weatherwise* magazine, 1948 to the present.

years past; it being generally held to be two feet [(60 cm)], or two feet and a half [(75 cm)], on the level, and in some places deeper."[6]

*27–28 January 1772.* "The Washington and Jefferson Snowstorm." George Washington in Mount Vernon and Thomas Jefferson in Monticello were marooned by this storm. Snowfall was estimated at 90 cm on a level across Virginia and Maryland. Washington wrote, ". . . the deepest snow which I suppose the oldest living ever remembers to have seen in this country."[7]

*26 December 1778.* "The Hessian Storm" was a severe blizzard accompanied by heavy snows, high winds, and bitter cold from Pennsylvania to New England, with drifts reported to 500 cm in Rhode Island. The storm was named for troops occupying Rhode Island during the Revolutionary War.

*28 December 1779–7 January 1780.* Three storms during one of the coldest winters of the past three centuries produced deep snows in much of New England. The first storm produced rain from New York City southward, but snow occurred in New England with 45 cm at New Haven, Connecticut. The second system was a violent snowstorm with extremely high tides from the Carolinas northward. The third storm was confined primarily to eastern New England. Snow depths in the wake of the three storms ranged between 60 and 120 cm from Pennsylvania to New England.

*4–10 December 1786.* This period featured another succession of three crippling snowstorms. Estimates in the local press placed snow depths from Pennsylvania to New England at 60 to 120 cm. In describing the third storm, a Boston newspaper notes, "a snowstorm equally severe and violent with that we experienced on Monday and Tuesday preceding. The quantity of snow is supposed to be greater, now, than has been seen in this country at any time since that which fell seventy years ago, commonly termed 'The Great Snow.' "[8]

*19–21 November 1798.* "The Long Storm" is described by Ludlum as the heaviest November snowstorm in the history of the coastal northeast from Maryland to Maine. Forty-five cm of snow reportedly fell in New York City.

## 2.2  NINETEENTH CENTURY

*26–28 January 1805.* This cyclone brought a very heavy snowstorm to New York City and New England. Snow fell continuously for 48 hours in New York City, where 60 cm reportedly accumulated.

*23–24 December 1811.* Temperatures fell from well above freezing on 23 December to −15°C on 24 December, while a storm intensified explosively off Long Island, New York. Snowfall in New York City, Long Island, and southern New England averaged 30 cm, as severe blizzard conditions prevailed. Strong winds and high tides caused extensive damage to shipping.

---

[6] Ludlum, D., 1966, p. 62.
[7] Ludlum, D., 1966, p. 145.
[8] Ludlum, D., 1966, p. 71.

*5–7 January 1821.* An extensive snowstorm spread from Virginia to southern New England, leaving 30 cm at Washington, D.C., 35 cm at Baltimore, Maryland, 45 cm at Philadelphia, Pennsylvania, and 35 cm in New York City.

*14–16 January 1831.* "The Great Snowstorm" produced the heaviest snowfall over the largest area of any storm studied by Ludlum. Accumulations exceeded 25 cm from the Ohio Valley across much of the Atlantic coast north of Georgia. Washington, D.C., reported 33 cm, with 45 cm at Baltimore, Maryland, 45 to 90 cm near Philadelphia, Pennsylvania, 37 to 50 cm at New York City, and 50 to 75 cm over southern New England.

*8–10 January 1836.* This event became known as "The Big Snow" for interior New York, northern Pennsylvania, and western New England, where 75 to 100 cm fell. The storm also buried the coastal plain, with 37 cm at Philadelphia, Pennsylvania, 37 to 45 cm at New York City, and 60 cm across southern New Jersey.

*18–19 January 1857.* "The Cold Storm" combined snowfall in excess of 25 cm with temperatures near or below $-15°C$ and high winds to produce severe blizzard conditions from North Carolina to Maine. Snow totals ranged from 37 cm near Norfolk, Virginia, to 45 to 60 cm in Washington, D.C., and 60 cm at Baltimore, Maryland, with 30 cm at New York City, and 35 cm at Boston, Massachusetts.

*11–14 March 1888.* The "Blizzard of '88" produced severe blizzard conditions over a 2- to 3-day period across Pennsylvania, New Jersey, New York, and central and western New England. Perhaps the most legendary of all historic snowstorms, it generated depths of 75 to 125 cm across sections of New York and New England, with extremely high winds and temperatures falling below $-15°C$. New York City was particularly hard hit, with widespread destruction of shipping and communications. See Kocin (1983) for a meteorological analysis of the storm and Werstein (1960) and Caplovich (1987) for descriptive and photographic accounts of the storm.

*26–27 November 1898.* The "Portland Storm" was a rapidly developing system that yielded record early-season snowfalls from the Middle Atlantic states to New England, accompanied by high winds. New York and Boston received about 25 cm, while New London, Connecticut, measured 68 cm. The event was named for the S.S. Portland, which sank offshore of Cape Cod during the storm.

*12–14 February 1899.* The "Blizzard of '99" formed along the leading edge of one of the greatest outbreaks of Arctic air ever experienced in the central and eastern United States. Snow fell from central Florida to Maine with 25- to 50-cm accumulations common from the Carolinas to New England. See Kocin et al. (1988) for a meteorological account of this episode.

## 2.3  TWENTIETH CENTURY (PRIOR TO 1955)

*25–26 December 1909.* The Christmas night storm produced heavy snow, high winds, and high tides throughout the Northeast. The snowfall of 53 cm in Philadelphia, Pennsylvania, was the greatest in modern records until 1983.

*1–2 March 1914.* This intense storm resulted in 30 to 60 cm of snow in Pennsylvania, New Jersey, and New York, along with high winds. The sea level pressure fell to 962 mb at New York City at the height of the storm, with near-hurricane-force winds and snow accumulations of 35 cm.

*3–4 April 1915.* A spring snowstorm produced accumulations of 25 to 50 cm throughout the Middle Atlantic states and southern New England with 48 cm at Philadelphia, Pennsylvania.

*4–7 February 1920.* This was a great snow and sleet storm that left an accumulation of 37 to 50 cm of ice, sleet, and snow in New York City and Boston, Massachusetts, stalling traffic for weeks.

*27–29 January 1922.* The "Knickerbocker Storm" affected the Middle Atlantic states on the 150th anniversary of the "Washington and Jefferson Storm." Exceptionally heavy snow fell in Virginia and Maryland. Seventy centimeters buried Washington, D.C., contributing to the collapse of the roof of the Knickerbocker Theatre where more than 100 people were killed.

*19–20 February 1934.* During the coldest February on record, this rapidly developing storm produced severe blizzard conditions throughout southern New England. This snowfall was one of Connecticut's worst in modern times, with 50-cm accumulations, strong winds, and temperatures that dropped from near 0°C to −15°C during the course of the storm.

*22–24 January 1935.* This was a widespread storm that left snow accumulations of 30 cm or more from Pennsylvania northward through New England.

*14–15 February 1940.* The "St. Valentine's Day Storm" was a rapidly developing system that paralyzed Boston and the rest of New England with snow depths exceeding 30 cm and very high winds.

*26–27 December 1947.* A surprise snowstorm brought 65 cm, the heaviest 24-hour accumulation of snow in New York City's modern records. Most of the snow actually fell within a 12-hour period. At White Plains, New York, 15 cm piled up in just 1 hour, with 49 cm over a 6-hour span.

During the years from 1955 through 1985, a number of winter storms have also attained historic stature in the Northeast. The blizzards of February 1958 and January 1966, the triple snowstorms of the 1960/1961 winter, the great New England wind and snowstorm of February 1978, the "Presidents' Day Storm" of February 1979, and the paralyzing urban storm of February 1983 are the most notable events of this period. These and other major snowstorms of this era will be the focus of the following chapters.

CHAPTER **3**

# Climatological Overview of the Period from 1955 to 1985

## 3.1 MEAN SEASONAL SNOWFALL

The mean distribution of seasonal snowfall over the northeastern corner of the United States is largely dependent upon latitude (Fig. 1), ranging from 15 cm in the southeastern corner of Virginia to greater than 250 cm across sections of central and northern New England, New York, and West Virginia. The increase in snowfall at higher latitudes is skewed from southwest to northeast by the Atlantic Ocean's moderating influence on temperatures near the coastline. The effect of elevation on temperature also shapes the distribution of snowfall, as demonstrated by the snowfall maxima over mountainous inland sections. The primary source of snowfall in this region is from midlatitude cyclonic weather systems. Areas adjacent to the Great Lakes and the highlands of West Virginia, western Pennsylvania, and Maryland, however, also receive significant snowfall from the passage of cold, continental air over the relatively warm surfaces of the Great Lakes, in concert with orographic effects.

## 3.2 CLIMATOLOGY OF HEAVY SNOW OCCURRENCES

The distributions of moderate and heavy snow events from 1955/1956 through 1984/1985 in each of five major cities that span the populous coastal region of the Northeast are summarized in Fig. 2. Both monthly and 30-year totals are presented for Boston, Massachusetts, New York, New York, Philadelphia, Pennsylvania, Baltimore, Maryland, and Washington, D.C., to provide background on the distribution and frequency of heavy snow events in this region.

During the 30-year sampling period, the total number of events that yielded greater than 10-cm accumulations ranges from 48 at Washington, D.C., to 65 in New York City, and 100 in Boston, Massachusetts. January and February account for the majority of these occurrences in Washington, Baltimore, Philadelphia, and New York, while Boston has a distinct maximum in January and a secondary peak in March. The months of December, January, February, and March account for nearly all of the events for each city.

The total number of events that produced at least 25-cm accumulations ranges

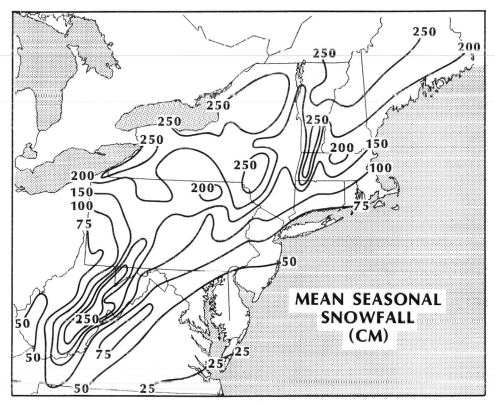

FIG. 1.   Mean seasonal snowfall [October through May (in cm)] for the northeastern United States.

from 8 at Washington, D.C., to 25 at Boston, Massachusetts. February displays the greatest frequency of heavy snow events in every city except Boston, where January has the largest number of heavy snowfalls. The frequency of 10- and 25-cm snow accumulations exhibit only small variations between Washington, D.C., Baltimore, Maryland, Philadelphia, Pennsylvania, and New York, New York. Washington recorded marginally fewer events and New York slightly more than the other cities. In contrast, Boston, Massachusetts, had a 50%–100% greater incidence of 10-cm snows and two to three times the number of 25-cm events than the other cities. Such a substantial difference indicates that Boston has a significantly greater potential for heavy snowstorms than any of the other cities.

To present a more general climatology of the coastal region between Virginia and Maine, 21 regions (Fig. 3) were selected to catalog the monthly and seasonal distributions of moderate and heavy snow occurrences for the winter seasons of 1955/1956 through 1984/1985. These regions are based on subdivisions found in *Climatological Data,* the monthly publication used to obtain the snowfall measurements. An occurrence of moderate (10-cm) or heavy (25-cm) snowfall is noted for each region in Fig. 3 whenever at least half the area received snow amounts in excess of the threshold values. This regional approach is used to develop a climatology of relatively widespread

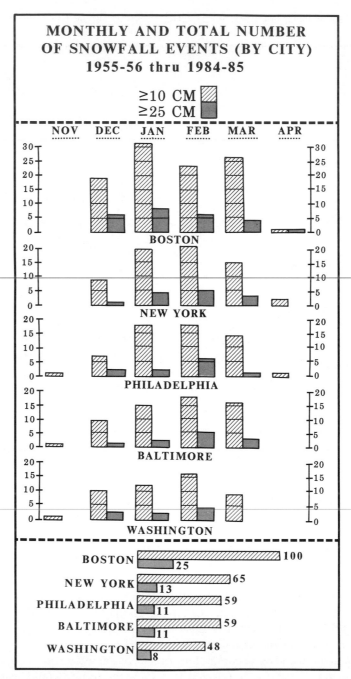

Fig. 2. Monthly and total number of snowfall events (by city) for the winter seasons 1955/1956 through 1984/1985 in Boston, Massachusetts, New York, New York, Philadelphia, Pennsylvania, Baltimore, Maryland, and Washington, D.C. Hatched shading represents the number of events exceeding 10 cm and dark shading represents the number of events exceeding 25 cm. Data were obtained from the following locations: Boston (Logan Airport); New York (La Guardia Airport, Central Park, Battery Place, and Kennedy Airport); Philadelphia (Philadelphia International Airport); Baltimore (Friendship or Baltimore–Washington International Airport); and Washington, D.C. (National Airport and Sterling, AFB). For cities with more than one snowfall measurement, an average value is used to designate "events."

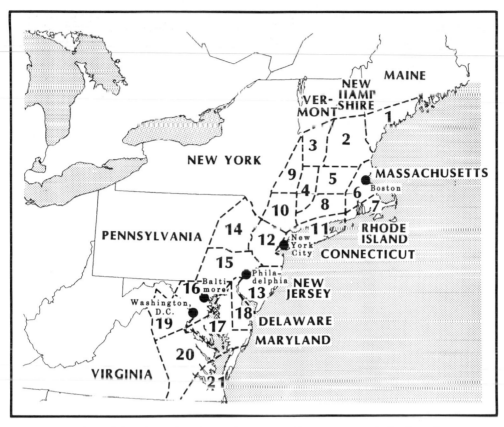

FIG. 3.    Locations of 21 regions selected to catalog the monthly and seasonal distributions of heavy snow occurrences described in Chapter 3.

snowfalls and to eliminate the isolated or localized moderate and heavy snow events that can occur occasionally [i.e., see Rosenblum and Sanders (1974)]. As a result, this method emphasizes those cases which have a significant impact on the greatest number of people. It is hoped that this approach will also remove irregularities caused by factors that make snowfall measurement an imprecise science. The presence of strong winds and differing sampling intervals often heighten measurement inaccuracies due to snow compaction and melting.

### 3.2.1    Snowfall Events Exceeding 10 cm

The number of storms that have produced snow accumulations exceeding 10 cm during the 30-year period from 1955/1956 through 1984/1985 (Table 1) ranges from 29 in southeastern Virginia (an average of 1 per winter season), to more than 200 in sections of central New England (an average of 7 events per season). Within the most densely populated regions of northern Virginia, Maryland, eastern Pennsylvania, New Jersey, extreme southeastern New York, and southern New England, a total of ap-

TABLE 1.   Total number of snowfall events exceeding 10 cm and 25 cm (by region) for the winter seasons 1955/1956 through 1984/1985, including the seasonal average of 10-cm events and the average time interval in years between 25-cm events.

| Region | Location | Total number of events exceeding 10 cm | Number of 10-cm events (seasonal average) | Total number of events exceeding 25 cm | Average time interval in years between 25-cm events |
|--------|----------|------|------|------|------|
| 1 | s Maine | 203 | 6.8 | 34 | 0.8 |
| 2 | s New Hampshire | 216 | 7.2 | 41 | 0.7 |
| 3 | s Vermont | 220 | 7.3 | 41 | 0.7 |
| 4 | w Massachusetts, nw Connecticut | 194 | 6.5 | 38 | 0.8 |
| 5 | c Massachusetts | 148 | 4.9 | 38 | 0.8 |
| 6 | e Massachusetts, n Rhode Island | 125 | 4.2 | 31 | 1.0 |
| 7 | se Massachusetts, s Rhode Island | 79 | 2.6 | 18 | 1.7 |
| 8 | c Connecticut | 103 | 3.4 | 19 | 1.6 |
| 9 | e New York | 160 | 5.3 | 33 | 0.9 |
| 10 | se New York | 109 | 3.6 | 24 | 1.3 |
| 11 | Long Island, New York City, s Connecticut | 75 | 2.5 | 15 | 2.0 |
| 12 | n New Jersey | 83 | 2.8 | 18 | 1.7 |
| 13 | s New Jersey | 51 | 1.7 | 11 | 2.7 |
| 14 | ne Pennsylvania | 100 | 3.3 | 23 | 1.3 |
| 15 | se Pennsylvania | 72 | 2.4 | 16 | 1.9 |
| 16 | c Maryland | 67 | 2.2 | 11 | 2.7 |
| 17 | se Maryland | 49 | 1.6 | 6 | 5.0 |
| 18 | Delaware | 50 | 1.7 | 7 | 4.3 |
| 19 | n Virginia | 63 | 2.1 | 15 | 2.0 |
| 20 | c Virginia | 53 | 1.8 | 7 | 4.3 |
| 21 | se Virginia | 29 | 1.0 | 3 | 10.0 |

proximately 50 to 80 events have occurred, averaging from slightly less than 2 to just under 3 occurrences per season. Further north and inland, the number of moderate and heavy snow events increases from approximately 100 to 150 across northeastern Pennsylvania, eastern New York, Connecticut, and Massachusetts, averaging 3 to 5 per season, to 200 or more over coastal Maine, southern Vermont, and New Hampshire, averaging about 7 events per season.

Storms capable of producing snowfall in excess of 10 cm typically occur during December, January, February, and March (Fig. 4). Accumulating snow has fallen as early as October and as late as May within the domain pictured in Fig. 4, but snowfall at these early and late dates is usually insignificant, and more likely to occur in northern and inland sections.

Within the period from 1955/1956 through 1984/1985, no region shown in Fig. 3 has experienced a general snowfall exceeding 10 cm in October. Accumulations of 10 cm or greater have been observed as early as 9 October (1979) in isolated areas of

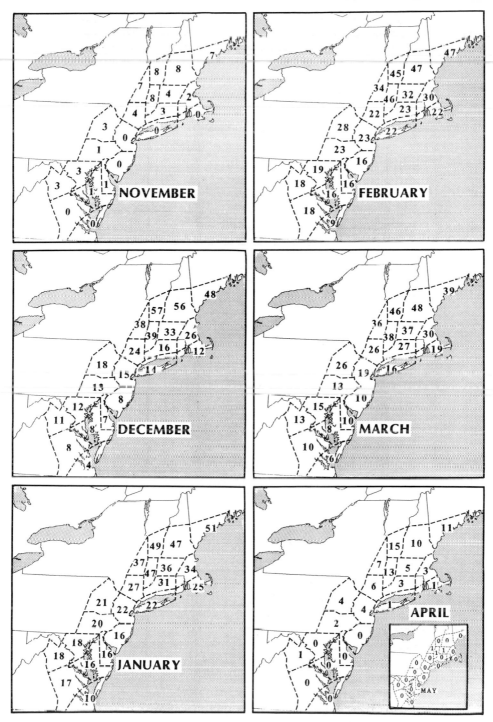

FIG. 4.    Number of snowfall events exceeding 10 cm (by region and month) for the winter seasons between 1955–56 and 1984–85.

TABLE 2.    Dates of snowstorms responsible for accumulations exceeding 25 cm over at least two regions defined in Fig. 3 (1955/1956 through 1984/1985). Underlined dates indicate the events comprising the 20 case analyses.

| Dates | No. of regions | Dates | No. of regions | Dates | No. of regions |
|---|---|---|---|---|---|
| 16–17 Mar 1956 | 7 | 10–11 Mar 1964 | 2 | 9–11 Jan 1974 | 5 |
| 18–20 Mar 1956 | 10 | 22–23 Jan 1966 | 5 | 5–6 Feb 1975 | 2 |
| 3–4 Dec 1957 | 3 | 26–27 Jan 1966 | 2 | 20–22 Dec 1975 | 5 |
| 7–8 Jan 1958 | 5 | 29–31 Jan 1966 | 10 | 7–18 Jan 1977 | 4 |
| 14–16 Jan 1958 | 2 | 25–26 Feb 1966 | 9 | 10–11 Jan 1977 | 2 |
| 14–17 Feb 1958 | 16 | 23–25 Dec 1966 | 8 | 6–7 Dec 1977 | 4 |
| 14–16 Mar 1958 | 6 | 5–7 Feb 1967 | 11 | 19–21 Jan 1978 | 13 |
| 18–21 Mar 1958 | 10 | 23 Feb 1967 | 4 | 5–7 Feb 1978 | 17 |
| 12–13 Mar 1959 | 6 | 6–7 Mar 1967 | 5 | 24–25 Dec 1978 | 3 |
| 29–30 Dec 1959 | 2 | 15–16 Mar 1967 | 4 | 25–26 Jan 1979 | 3 |
| 19–20 Feb 1960 | 2 | 22–23 Mar 1967 | 2 | 18–20 Feb 1979 | 8 |
| 2–5 Mar 1960 | 17 | 23–29 Dec 1967 | 6 | 1–2 Mar 1980 | 2 |
| 10–13 Dec 1960 | 15 | 8–10 Feb 1969 | 11 | 6–7 Dec 1981 | 3 |
| 18–20 Jan 1961 | 11 | 22–28 Feb 1969 | 6 | 15–16 Dec 1981 | 4 |
| 2–5 Feb 1961 | 14 | 22–23 Dec 1969 | 2 | 13–15 Jan 1982 | 2 |
| 20–21 Nov 1961 | 2 | 25–28 Dec 1969 | 10 | 5–7 Apr 1982 | 11 |
| 14–15 Feb 1962 | 7 | 22–24 Dec 1970 | 2 | 15–16 Jan 1983 | 5 |
| 6–7 Mar 1962 | 2 | 3–4 Mar 1971 | 4 | 7–8 Feb 1983 | 7 |
| 23–24 Jan 1963 | 2 | 20–21 Mar 1971 | 3 | 10–12 Feb 1983 | 16 |
| 25–26 Jan 1963 | 2 | 25–26 Nov 1971 | 6 | 10–11 Jan 1984 | 2 |
| 11–14 Jan 1964 | 14 | 18–20 Feb 1972 | 12 | 13–14 Mar 1984 | 6 |
| 16 Feb 1964 | 3 | 15–16 Mar 1972 | 4 | 28–29 Mar 1984 | 3 |
| 19–20 Feb 1964 | 3 | 29–30 Jan 1973 | 4 | | |

the Middle Atlantic states and, following the 30-year period of study, as early as 3–4 October (1987) across eastern New York and western New England. In November, episodes of general snowfall exceeding 10 cm occur sporadically, with a frequency as great as once every 3 to 4 years over inland sections of southern and central New England. After our 30-year subject period, an exceptional early-season snowstorm swept across portions of the Northeast coastal region on 10–12 November 1987, with accumulations reaching 25 to 35 cm in the Washington, D.C., and Boston, Massachusetts, areas. By December, all regions can experience snow events exceeding 10 cm. The far northern section of the domain receives moderate to heavy snow events with equal likelihood in any of the months from December through March. Across extreme southern New England and much of the coastal region of the Middle Atlantic states, the frequency of moderate to heavy snowfalls reaches a peak in January and February (Fig. 4). From Virginia to southern New England, the frequency of moderate to heavy snow events diminishes in March, although there is a slightly greater tendency for snowstorms to occur in this month than in December. In all areas, the incidence of winter-type storms declines quickly after 1 April. An unusually heavy snowfall was observed over portions of interior New York state and southern New England on 9 May 1977, the latest such occurrence in the sample.

TABLE 3.    Dates of 20 selected Northeast coast snowstorms and snowfall amounts (cm) for Washington, D.C. (the maximum amount at either Dulles International Airport or National Airport), Baltimore, Maryland, Philadelphia, Pennsylvania, New York, New York (the maximum amount at either Central Park, LaGuardia Airport, Kennedy International Airport, or Battery Place), and Boston, Massachusetts. Unofficial estimates of local snowfall amounts are noted in parentheses for the storm of March 1958.

| | Snowfall in cm | | | | |
|---|---|---|---|---|---|
| Storm date | Washington, D.C. | Baltimore, Maryland | Philadelphia, Pennsylvania | New York, New York | Boston, Massachusetts |
| 18–20 Mar 1956 | 4 | 14 | 22 | 34 | 34 |
| 14–17 Feb 1958 | 37 | 39 | 33 | 26 | 49 |
| 18–21 Mar 1958 | 12 (50+) | 21 (50+) | 29 (50+) | 30 (50+) | 15 |
| 2–5 Mar 1960 | 20 | 26 | 21 | 39 | 50 |
| 10–13 Dec 1960 | 22 | 36 | 37 | 44 | 33 |
| 18–20 Jan 1961 | 20 | 21 | 34 | 25 | 31 |
| 2–5 Feb 1961 | 21 | 27 | 26 | 61 | 37 |
| 11–14 Jan 1964 | 26 | 25 | 18 | 32 | 23 |
| 29–31 Jan 1966 | 35 | 31 | 21 | 17 | 16 |
| 23–25 Dec 1966 | 23 | 22 | 32 | 18 | 14 |
| 5–7 Feb 1967 | 30 | 27 | 25 | 39 | 24 |
| 8 10 Feb 1969 | 13 | 8 | 7 | 51 | 28 |
| 22–28 Feb 1969 | — | — | 5 | 4 | 67 |
| 25–28 Dec 1969 | 31 | 15 | 13 | 19 | 11 |
| 18–20 Feb 1972 | 25 | 8 | 9 | 16 | 16 |
| 19–21 Jan 1978 | 19 | 14 | 34 | 36 | 54 |
| 5–7 Feb 1978 | 6 | 23 | 36 | 45 | 69 |
| 18–20 Feb 1979 | 47 | 51 | 36 | 32 | — |
| 5–7 Apr 1982 | — | — | 9 | 24 | 34 |
| 10–12 Feb 1983 | 58 | 58 | 54 | 56 | 34 |

## 3.2.2    Snowfall Events Exceeding 25 cm

Storms that deposit 25 cm or more are relatively rare (Table 1). Only three such storms occurred over the 30-year period in southeastern Virginia, an average of one every 10 years, with 5 to 11 events over the coastal regions from Virginia to southern New Jersey, approximately one every 3 to 5 years. The heavily populated urban corridor from northern Virginia through extreme southern New England experienced 11 to 18 major snowfalls, approximately one every other year. The inland areas of northeastern Pennsylvania, eastern New York, and southern New England received 19 to 33 heavy snows, approximately one every 1 to 2 years, while 34 to 41 events were noted across the northern limits of the domain, slightly greater than one every year.

These major snowstorms typically occur from December through March (Fig. 5), with very few events in November or April. For much of southern New England and the Middle Atlantic states, the frequency of storms with snowfall exceeding 25 cm shows a significant maximum during February. This contrasts with the nearly equal likelihood of 10-cm amounts in January and February (Fig. 4). Thus, the numbers shown in Fig. 5 support the notion that February is the "big month" for the "big

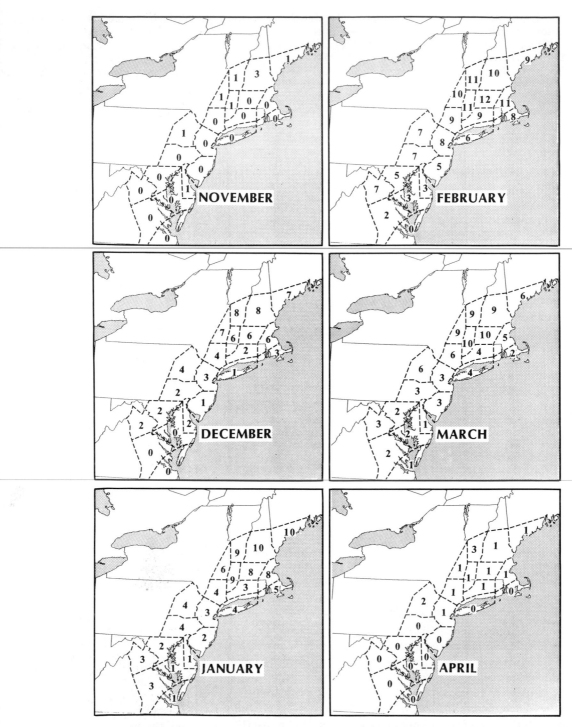

FIG. 5.    Number of snowfall events exceeding 25 cm (by region and month) for the winter seasons between 1955–56 and 1984–85.

snow" along the Northeast coast, at least during the 30-year period between 1955 and 1985.

The period from 1955/1956 through 1984/1985 contains a total of 68 storms that produced snowfalls of 25 cm or more in at least 2 of the 21 sections shown in Fig. 3. The dates of these storms and the number of sections with accumulations exceeding 25 cm are provided for each case in Table 2. Twenty of the 68 cases, each underlined in Table 2, are the subject of the following two chapters, where the general characteristics and structure of the most paralyzing storms are examined in depth. The 20 cases are selected because they had the greatest impact on the largest number of people within the domain that spans the coastal region from Virginia to southern Maine, including the metropolitan centers of Washington, D.C., Baltimore, Maryland, Philadelphia, Pennsylvania, New York, New York, and Boston, Massachusetts. While some of the other cases in Table 2 affected almost as many regions as the 20 cases selected (e.g., 16–17 March 1956, 14–16 March 1958, 12–13 March 1959, 14–15 February 1962, 25–26 February 1966, 28–29 December 1967, 25–26 November 1971, 7–8 February 1983, and 13–14 March 1984), they were omitted because the heaviest accumulations occurred across less populated areas.

The total snowfall at Washington, D.C., Baltimore, Maryland, Philadelphia, Pennsylvania, New York, New York, and Boston, Massachusetts, for each of the 20 storms is listed in Table 3. The 20 selected cases encompass many of these cities' greatest snowfalls over the 30-year period. Our collection includes 7 of the 8 storms that produced accumulations exceeding 25 cm in Washington, D.C., 10 of 11 in Baltimore, Maryland, 11 of 11 in Philadelphia, Pennsylvania, 13 of 13 in New York, New York, and 13 of the 25 events in Boston, Massachusetts. Snow connoisseurs from Washington, Baltimore, Philadelphia, and New York should be pleased with the storm selection presented here, while Boston fans might be somewhat disappointed that more cases were not included.

CHAPTER **4**

# Descriptions of 20 Major Snowstorms

In this chapter, patterns of snowfall, sea level pressure, surface wind, temperature, and other parameters are examined for the ensemble of 20 cases. These patterns were derived from a survey of climatological data and surface charts developed from National Meteorological Center (NMC) analyses available on microfilm from the National Climatic Data Center (NCDC) in Asheville, North Carolina. The goals of this study are to identify the meteorological patterns that characterize major snow events, to determine what case-to-case variability might make generalizations difficult to apply, and to describe the dynamical and physical processes associated with these patterns. A description of upper-level meteorological fields will be presented in the next chapter.

## 4.1 SNOWFALL DISTRIBUTION OF 20 STORMS

The distribution of heavy snowfall for each of the 20 cases is shown in Fig. 6. The shaded regions, which depict amounts exceeding 25, 50, and 75 cm, show that heavy snow accumulations occur in bands averaging 200 to 500 km in width which cover distances of 500 km or more and are generally oriented from southwest to northeast. Snowfalls of 50 cm or greater were observed in every storm, with the most widespread distributions displayed by the cases of 22–28 February 1969, 25–28 December 1969, 5–7 February 1978, and 10–12 February 1983. Accumulations exceeding 75 cm were realized in eight of the cases, and were most common in eastern New England during the slow-moving storm of late February 1969.

The widely varying local distributions of heavy snowfall seen in Fig. 6 point out the difficulties inherent in predicting snow amounts, as illustrated by the examples which follow. The storm of 9–10 February 1969 produced snow accumulations of greater than 50 cm in the northwestern suburbs of Boston, and 25 to 50 cm within the city limits, while insignificant amounts occurred in the southeastern suburbs, where the snow was mixed with rain and ice pellets. The "Presidents' Day Storm" of February 1979 brought snowfalls of up to 40 cm across northeastern New Jersey and New York City, while the northern suburbs measured 10 cm or less, and much of Connecticut and southern New England received no snow at all. During other episodes, such as the storms of January and February 1961, the greatest snow amounts fell along the northern fringe of the band of heavy snowfall. In the storm of March 1958, differences

## STORM SNOWFALL FOR TWENTY CASES
### ( ≥ 25 CM )

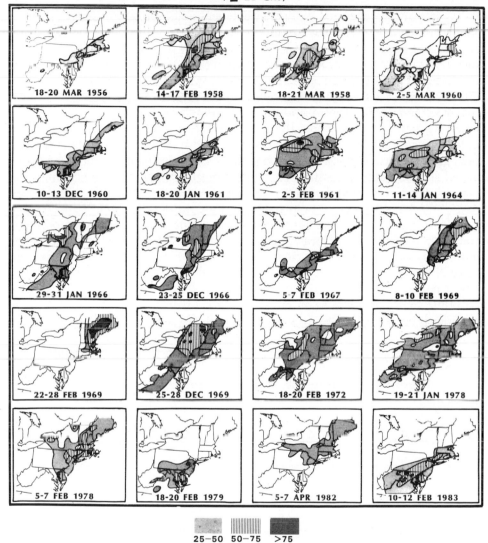

**25–50    50–75    >75**

FIG. 6.    Storm snowfall in excess of 25 cm for 20 snowstorms between 1955 and 1985 (light shading, 25–50 cm; linear shading, 50–75 cm; heavy shading, greater than 75 cm).

in elevation of only 100 to 200 m were the critical factor between mixed precipitation and snowfalls that exceeded 50 cm in parts of Pennsylvania and Maryland. Such extreme variability within a small area must have presented a tremendous challenge to local forecasters! The physical and dynamical processes that affect the spatial distribution and intensity of the snowfall associated with these storms are addressed in the following sections and chapters.

## 4.2  CYCLONES

All 20 cases are characterized by the development and propagation of well-defined cyclonic circulations, marked by a minimum of sea level pressure along the East Coast of the United States. The general characteristics of low-pressure centers that develop along the East Coast have been described previously by Austin (1941), Miller (1946), and Mather et al. (1964). Miller examined 200 cases of varying intensity spread over a 10-year period ending in 1939 and categorized East Coast storm development into two types, "A" and "B." Type A storms resemble the classical polar front wave cyclone studied by the Norwegian school (Bjerknes 1919; Bjerknes and Solberg 1922). The surface low, a minimum of sea level pressure defined by the formation of at least one closed isobar, develops along a frontal boundary separating an outbreak of cold continental air from warmer maritime air. Type-B storms represent a more complex sea level cyclonic development that Miller describes as a phenomenon unique to the East Coast of the United States. Type-B systems feature the development of a secondary area of low pressure along the East Coast to the southeast of an occluding primary, or initial, low-pressure center located over the Ohio Valley. The secondary low-pressure center forms along the primary cyclone's warm front, which separates a shallow wedge of cold air between the Appalachian Mountains and the Atlantic Ocean from warmer air over the ocean.

While some doubt exists concerning a physical basis for the "A" and "B" classifications proposed by Miller, his study cites secondary sea level development as one of the most intriguing aspects of cyclogenesis along the East Coast of the United States. The formation of a secondary low-pressure center is of particular concern to forecasters, since their ability to predict the onset of cyclogenesis and the subsequent track of the surface center is a key factor in the forecast of when and where the heaviest snows will occur.

### 4.2.1  Tracks and Development

Paths of the surface low-pressure centers associated with the 20 selected cases are shown in Fig. 7. The cases are grouped to distinguish between systems exhibiting distinct secondary low-pressure centers along the East Coast (Figs. 7a and 7b) and those that do not display secondary development (Figs. 7c and 7d).

Ten of the 20 cases exhibited the type-B scenario described by Miller (Figs. 7a and 7b; Table 4), in which a distinct second low-pressure center formed to the southeast of a preexisting primary low, as depicted schematically in Fig. 8a1. The primary low-pressure centers followed diverse paths that originated over the western Gulf of Mexico, the southern Plains states, and the upper Midwest, with a locus of tracks oriented from the southern Plains states to the upper Tennessee and Ohio valleys. As each initial low-pressure center moved toward the Appalachian Mountains, a secondary center developed over the northeastern Gulf of Mexico, Georgia, the Carolinas, or the offshore waters. Once the secondary low had formed, the primary low then dissipated. The secondary low-pressure centers followed similar paths from eastern North Carolina northeastward to the waters offshore of the Middle Atlantic states and southern New England, passing within approximately 100 to 300 km of the coast.

# PATHS OF CYCLONES AT SEA LEVEL

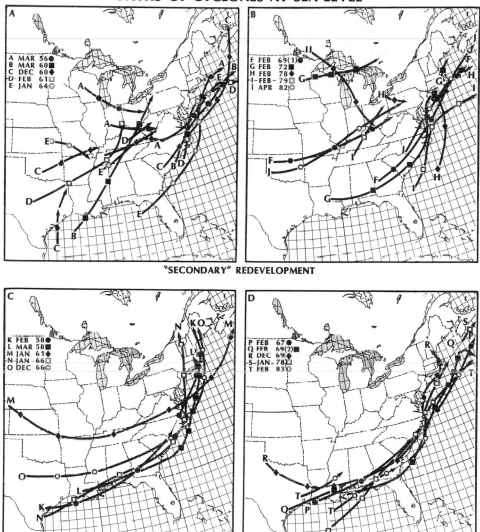

"SECONDARY" REDEVELOPMENT

"PRIMARY" OR NO REDEVELOPMENT

FIG. 7. Paths of low pressure centers at sea level for 20 cases. Paths are grouped according to (a, b) cases exhibiting "secondary" redevelopment and (c, d) cases exhibiting no redevelopment or only "primary" redevelopment ("center jumps"). Symbols along path represent 12-hourly positions.

The distinctive separation of paths that characterizes the cases with secondary cyclogenesis is not observed in the other 10 storms (Figs. 7c and 7d; Table 4). All but one of these cases developed in or propagated across the Gulf Coast region. These systems then passed south of the Appalachian Mountains, and headed northeastward along the East Coast, near or offshore of the Middle Atlantic and southern New England coasts. The low-pressure center associated with the January 1961 "Kennedy Inaugural

**TABLE 4.  Characteristics of cyclonic development at sea level, including redevelopment characteristics, average propagation rates over the 60-hour study period (P for primary low, S for secondary low), lowest central sea level pressures analyzed (mb), duration (hours), and amount (mb) of deepening [≥ −1 mb (3 h)⁻¹], duration (hours) of rapid deepening [≥ −3 mb (3 h)⁻¹; P for primary low, S for secondary low], and the duration (hours) of the primary low once the secondary low-pressure center had formed.**

| Date | Redevelopment characteristics | Average propagation rate (m s⁻¹) over 60-hour study period | Lowest central sea level pressure analyzed (mb) | Duration of deepening (h) [≥ −1 mb (3 h)⁻¹] | Amount of deepening (mb) [≥ −1 mb (3 h)⁻¹] | Duration of rapid deepening (h) [≥ −3 mb (3 h)⁻¹] | Duration of primary low (hours) following formation of secondary low center |
|---|---|---|---|---|---|---|---|
| 18–20 Mar 1956 | secondary | 9 (P) 11 (S) | 1000 | 21 | 11 | 0 (P)  0 (S) | 6 |
| 14–17 Feb 1958 | center jump | 17 | 970 | 36 | 36 | 21 | |
| 18–21 Mar 1958 | none | 6 | 976 | 45 | 33 | 18 | |
| 2–5 Mar 1960 | secondary | 18 (P) 12 (S) | 960 | 42 | 52 | 3 (P) 24 (S) | 15 |
| 10–13 Dec 1960 | secondary | 15 (P) 17 (S) | 966 | 39 | 44 | 3 (P) 21 (S) | 6 |
| 18–20 Jan 1961 | none | 19 | 964 | 48 | 52 | 24 | |
| 2–5 Feb 1961 | secondary | 13 (P) 12 (S) | 988 | 33 | 30 | 0 (P) 12 (S) | 9 |
| 11–14 Jan 1964 | secondary | 11 (P) 11 (S) | 982 | 45 | 30 | 3 (P) 12 (S) | 24 |
| 29–31 Jan 1966 | center jump | 20 | 970 | 33 | 42 | 27 | |
| 23–25 Dec 1966 | center jump | 15 | 978 | 42 | 31 | 18 | |
| 5–7 Feb 1967 | center jump | 24 | 976 | 30 | 29 | 15 | |
| 8–10 Feb 1969 | secondary | 14 (P) 14 (S) | 966 | 42 | 39 | 0 (P) 18 (S) | 9 |
| 22–28 Feb 1969 | none | 14 | 993 | 27 | 23 | 12 | |
| 25–28 Dec 1969 | center jump | 16 | 976 | 36 | 33 | 18 | 27 |
| 18–20 Feb 1972 | secondary | 9 (P) 12 (S) | 973 | 33 | 36 | 0 (P) 12 (S) | |
| 19–21 Jan 1978 | center jump | 17 | 995 | 30 | 19 | 6 | |
| 5–7 Feb 1978 | secondary | 12 (P)  9 (S) | 984 | 39 | 42 | 0 (P) 15 (S) | 15 |
| 18–20 Feb 1979 | secondary | 10 (P) 15 (S) | 995 | 27 | 24 | 0 (P) 12 (S) | 15** |
| 5–7 Apr 1982 | secondary | 19 (P) 15 (S) | 966 | 33 | 36 | 6 (P) 18 (S) | 6 |
| 10–12 Feb 1983 | center jump | 14 | 994 | 30 | 18* | 6 | |

* Deepening continued after study period ended

** Primary and secondary low centers developed simultaneously; inverted trough associated with primary low preceded the coastal trough associated with secondary cyclogenesis

### CYCLONIC REDEVELOPMENT AT SEA-LEVEL

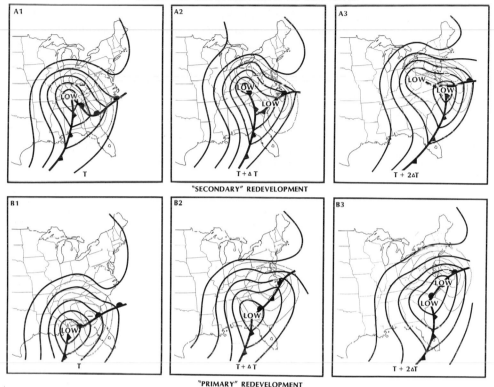

FIG. 8.   Schematic representation of cyclogenesis along the East Coast, including (a) "secondary" redevelopment as an Ohio Valley system fills and (b) "primary" redevelopment or center jump as a Gulf of Mexico system moves northeastward. Solid lines are isobars and thin dotted lines represent isallobars for regions of greatest pressure falls.

Storm" is the one exception, since it followed a path (labeled "M" in Fig. 7c) more typical of cyclones that undergo secondary redevelopment. It is possible that the rapid propagation of this system masked a redevelopment of the sea level center in south-eastern Virginia while the primary low dissipated over the Tennessee–North Carolina mountains, as inferred from 3-hourly NMC analyses.

Although a secondary low center was not observed in these 10 cases, 7 of the 10 (see Table 4) exhibited a center jump. In cases displaying a center jump, the surface low-pressure center propagated northeastward along the Southeast coast and appeared to suddenly redevelop further northeast along the coast (Fig. 8b). This redevelopment occurs along the same path as that of the primary low center, in contrast to the separate tracks that mark secondary cyclogenesis (Fig. 8a), and are therefore labeled "primary redevelopment" in Fig. 8. The sudden acceleration of the surface low that characterized center jumps, or primary redevelopments, occurred almost exclusively along or near the coast of the Carolinas (see dotted tracks in Figs. 7c and 7d).

A comparison of the coastal cyclone tracks from all of the cases shown in Fig. 7

with the snowfall distributions shown in Fig. 6 indicates that the axis of heaviest accumulations is usually found 100 to 300 km to the left of, and roughly parallel to, the paths of low-pressure centers that track northeastward up the Atlantic coast.

In total, 17 of the 20 cases clearly show a redevelopment and/or sudden acceleration of the surface low along the East Coast, or both. The clustering of cyclone tracks near the coastline in systems that underwent secondary cyclogenesis or a center jump indicates that these storms are influenced by the underlying topography. The Appalachian Mountains appear to mark a transition between the areas where primary storm systems weaken and the region in which secondary storms develop and intensify farther east. Secondary cyclogenesis and center jumps also seem to be confined to the vicinity of the Carolina coast, suggesting that the coastline itself plays an important role.

Petterssen (1956, p. 267) showed that cyclogenesis occurs primarily in association with mountain chains and coastlines, especially near the Rocky Mountains, the Alps, the Himalayas, and along the eastern coasts of North America and Asia. A mechanism termed "lee cyclogenesis" relates a proclivity for cyclonic development on the leeward or downwind sides of mountain chains to changes in vorticity when vortices are "stretched" as they cross mountains [i.e., see Holton (1979, pp. 87–89) and Smith (1979, pp. 163–169)]. This mechanism appears to be most important in the lee of extensive and steep mountain chains, such as the Rocky Mountains (i.e., Newton 1956; Carlson 1961; Egger 1974; and others) and the Alps (i.e., Bleck 1977; Buzzi and Tibaldi 1978; Smith 1984; and others). The influence of the Appalachian Mountains on cyclogenesis is presently not well known. The relatively small horizontal and vertical dimensions of the Appalachians suggest, however, that their influence may be much less than that of more prominent mountain ranges. Nevertheless, other processes related to the presence of the mountains are important factors in maintaining cold air over the coastal plain, and contribute to the enhanced low-level baroclinicity required for cyclogenesis (see section 4.3.1).

The coastline itself serves as a focal point for cyclogenesis since it represents the natural boundary between cold air trapped over the coastal plain and boundary-layer air warmed and moistened by the Atlantic Ocean. The coastline also represents a discontinuity in surface roughness that influences low-level divergence and enhances cyclogenetic circulation patterns (Bosart 1975; Anthes and Keyser 1979; Ballentine 1980; Danard and Ellenton 1980). The shape of the coastline also plays a role in the cyclogenetic process, since concave coastlines tend to promote cyclogenesis (Petterssen 1956; Godev 1971a,b; Bosart 1975; Ballentine 1980). The North Carolina and New England coasts exhibit a concave shape and are two areas prone to cyclogenesis and coastal frontogenesis (Bosart 1975). The occurrence of center jumps along the coast of the Carolinas may be indirectly related to this factor.

### 4.2.2   Propagation Rates

The 12-hourly positions of the sea level low-pressure centers (Fig. 7) illustrate considerable case-to-case variability in their propagation rates, which is related to case-to-case differences in snowfall amounts. Mean propagation rates for each case were computed from 3-hourly low-pressure positions analyzed by NMC and are provided in Table 4.

The surface low-pressure center associated with the snowstorm of February 1967 propagated at a mean rate of 24 m s$^{-1}$ from the Gulf Coast to east of Nova Scotia. Snow amounts along the axis of heaviest falls were generally in the range of 25–30 cm, as snow fell for a period of 15 hours or less. In contrast, the surface low-pressure center associated with the snowstorm of March 1958 progressed northeastward at an unusually slow rate, averaging only 6 m s$^{-1}$ over a 60-hour period. During this storm, snow fell continuously for 3 days at some locations, with accumulations exceeding 75 cm.

A closer examination of the individual cyclone paths in Fig. 7 reveals that most cases are marked by significant changes in speed. In 13 of the 20 storms, the surface low reduced its average speed between 12-hour time periods by at least 5 m s$^{-1}$ along the Northeast coast. This "stalling" of the sea level center occurred most frequently from east of the New Jersey coast to southern New England, and its most obvious effect was to prolong snowfall, especially in New England. In the storms of March 1956, March 1960, February 1969, December 1969, and February 1978, the surface lows stalled after periods of faster progression. In each case, the center moved 250 km or less over a 12-hour period, at a rate of less than 8 m s$^{-1}$, resulting in excessive snowfall. The deceleration occurs as a manifestation of the occlusion process when the surface and upper-level low centers become nearly collocated. This factor poses a considerable challenge to forecasters concerned with predicting an end to the snowfall and the final accumulation.

The most notorious example of a stalled cyclone is the March 1888 "Blizzard of '88." This storm initially raced northeastward from the Carolina coast to southern New England, then remained nearly stationary close to the southern New England coast for 2 days. A protracted period of very heavy snow, strong winds, and extremely cold temperatures resulted (Kocin 1983).

### 4.2.3 Deepening Rates

Deepening rates were examined as a measure of the intensity of the sea-level cyclones associated with the major snowstorms. To describe the deepening rates, the central sea-level pressures were plotted at 3-hour intervals for a 60-hour period that includes the initiation, development, and decay of each storm[9] (Fig. 9). Central pressures were plotted for the cases exhibiting secondary redevelopment, including both the primary and secondary low centers (Figs. 9a1 and 9a2), and for cases in which no secondary development was observed (Figs. 9b1 and 9b2), as described in section 4.2.1. Table 4 lists the minimum sea-level pressure of each storm during the 60-hour period and summarizes the deepening characteristics of all 20 cases, including the duration of rapid intensification.

The most obvious generalization revealed by Fig. 9 is the significant decrease of central sea-level pressure in nearly every case, although the lowest pressure attained

[9] The 60-hour period also coincides with the time interval during which 12-hourly weather analyses, consistent with the temporal resolution of the operational rawinsonde network, are presented in Chapters 7 and 8. It should also be noted that there is considerable uncertainty regarding the accuracy of the sea-level pressure analyses of many storms over the Atlantic Ocean, due to the limited coverage of operational data.

## MINIMUM SEA-LEVEL PRESSURES

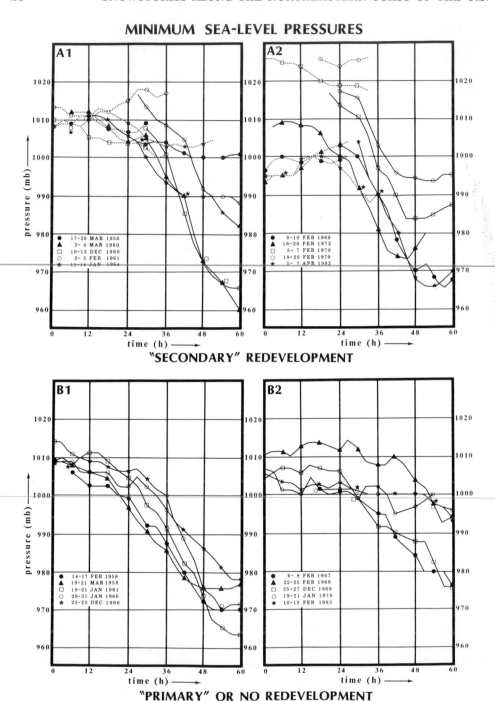

FIG. 9.   Time series of central pressures of the sea-level cyclones over the 60-hour study period. The 20 cases are grouped as in Fig. 7. In the top two panels, the pressures of the primary low are represented by the thin dotted lines and the secondary low by the solid lines.

by each storm is quite variable. As shown in Table 4, the absolute minimum sea level pressures for the 20 cases range from 960 mb (4 March 1960) to 1000 mb (18 March 1956), with an average slightly less than 979 mb.

Sea-level deepening, defined as a decrease in central pressure of at least $-1$ mb $(3 \text{ h})^{-1}$, generally occurred over a 24- to 48-hour period for all 20 storms.[10] The total amount of deepening exhibited in the sample varied from a minimum of 11 mb (18–19 March 1956) to a maximum of 52 mb (2–4 March 1960; 18–20 January 1961). The average period of deepening lasted approximately 36 hours, with a mean decrease in central sea level pressure of 33 mb. These values are consistent with Sanders' (1986a) observations of the duration and amount of deepening in a large sample of rapidly intensifying cyclones over the west-central North Atlantic Ocean. Most of the cases here contained a clearly defined and continuous period of intensification. Only a few storms [e.g., January 1964 and December 1969 (Fig. 9)] displayed multiple periods of deepening, separated by significant intervals of little or no change in central sea level pressure.

Sanders and Gyakum (1980) and Roebber (1984) found that explosive cyclogenesis is a common phenomenon over the waters off the East Coast of the United States [cf. Fig. 3 in Sanders and Gyakum (1980)]. Rapid development, defined as a deepening rate exceeding $-3$ mb $(3 \text{ h})^{-1}$, occurred over periods of 6 to 27 hours in all but one case (March 1956), which exhibited no 3-hour period of rapid deepening. Large deepening rates were observed in the other 19 storms, primarily along the coasts of the Carolinas, the Middle Atlantic states, and east-northeastward over the Atlantic Ocean (Fig. 10). The tracks of these storms are situated closer to the East Coast than the locus of rapidly developing cyclone tracks described by Sanders [1986a (cf. his Fig. 2)] for a 4-year sample of rapidly deepening cyclones over the western Atlantic Ocean.

In the 10 cases marked by secondary redevelopment (see Table 4), rapid deepening of the *primary* low-pressure center was observed in only 4 cases and lasted no longer than 6 hours. Meanwhile, nearly all of the *secondary* lows underwent a 12- to 24-hour period of rapid deepening. The only exception is the 18–20 March 1956 system, in which rapid intensification was not observed with the development of the secondary low-pressure center. As a rule, rapid deepening commenced shortly following the formation of the secondary cyclone center. The primary low center either filled (pressure rose) or was absorbed into the expanding circulation of the secondary low within 6 to 27 hours following the secondary cyclogenesis (Table 4).

Among the 10 storms that either exhibited a center jump or no redevelopment, deepening rates exceeding $-3$ mb $(3 \text{ h})^{-1}$ occurred in *all* cases, for periods ranging from 6 to 27 hours (Table 4).

A comparison of the deepening rates displayed in Fig. 9 and the precipitation analyses shown for each case in Chapter 7 indicates that moderate to heavy precipitation and snowfall were commonly observed along the Northeast coast when rapid intensification was in progress, resulting from the enhanced moisture and thermal advections that accompany explosive deepening. During the February 1979 "Presidents' Day

---

[10] The 3-hour interval is the resolution of NMC surface analyses, from which Fig. 9 is based.

# CYCLONE PATHS DURING PERIODS OF RAPID INTENSIFICATION

## (>-3 mb (3h)$^{-1}$)

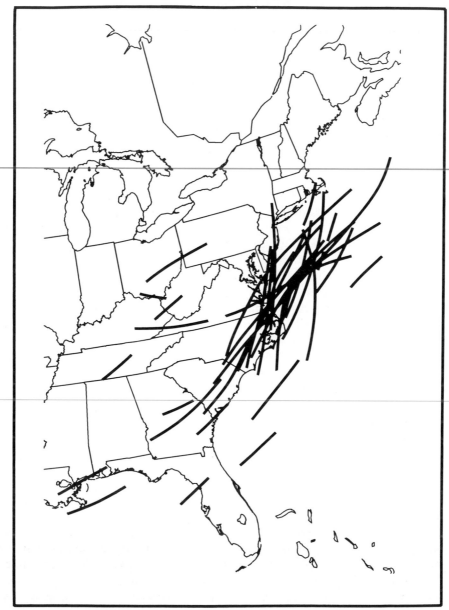

FIG. 10.    Paths of all low-pressure centers during periods of rapid intensification [sea-level pressure fell at $-3$ mb $(3$ h$)^{-1}$ or greater].

TABLE 5. Measures of sea-level pressure gradients surrounding the low-pressure center (see text for details), including the percent change in maximum pressure gradient over a 24-hour period defined in the text, a ranking of the maximum analyzed gradient, with respect to the other cases during the period, and the direction of the maximum gradient relative to the center.

| Date | Percent change in maximum gradient over 24-hour period | Ranking of maximum gradient | Direction of maximum gradient |
|------|-----|-----|-----|
| 18–20 Mar 1956 | +118 | 15 | NE |
| 14–17 Mar 1958 | +150 | (10) | NE |
| 18–21 Mar 1958 | +68 | 9 | NE |
| 2–5 Mar 1960 | +150 | 6 | NE |
| 10–13 Dec 1960 | +169 | 1 | N |
| 18–20 Jan 1961 | +183 | (7) | N |
| 2–5 Feb 1961 | +200 | 4 | NE |
| 11–14 Jan 1964 | +75 | 12 | NE |
| 29–31 Jan 1966 | +65 | 8 | W |
| 23–25 Dec 1966 | +107 | 11 | NE |
| 5–7 Feb 1967 | +131 | (10) | NW |
| 8–10 Feb 1969 | +280 | 3 | NE–N–NW |
| 22–28 Feb 1969 | +171 | 17 | NE |
| 25–28 Dec 1969 | +122 | (13) | NE |
| 18–20 Feb 1972 | +189 | 14 | NE |
| 19–21 Jan 1978 | +43 | 16 | NE |
| 5–7 Feb 1978 | +457 | 2 | NE |
| 18–20 Feb 1979 | +200 | (7) | W |
| 5–7 Apr 1982 | +500 | 5 | W |
| 10–12 Feb 1983 | +237 | (13) | NE |

Storm," snowfall rates increased to 5–10 cm h$^{-1}$ from Washington, D.C., to New York City early on the morning of 19 February, coinciding with the onset of strong sea-level deepening (see the 24- to 36-hour interval in Fig. 9a2). The storms of March 1956 and January 1978 differed from the other cases, however, in that intense sea-level development did *not* accompany the heavy snowfalls. So, while rapid sea-level development is a common characteristic of many of these storms, it is not a prerequisite for widespread heavy snowfall in the coastal sections of the Northeast.

Explosive cyclonic development over the western Atlantic Ocean is of considerable interest since many of the major storms examined here produced much of their heavy snowfall while undergoing rapid cyclonic intensification, and because of the severe impact that these storms have in offshore fishing and shipping operations. Factors contributing to the enhanced deepening rates include upper-level forcing associated with preexisting upper-level troughs and their associated patterns of divergence and vorticity advections (Sanders and Gyakum 1980; Sanders 1986a, Rogers and Bosart 1986), jet-streak circulation patterns and their contributions to vertical motions and vertical transports of stratospheric air (Uccellini et al. 1985; Uccellini 1986; Uccellini and Kocin 1987; Whitaker et al. 1988), and possible interactions of upper- and lower-level potential vorticity maxima during the period of rapid development (Gyakum

1983a; Farrell 1984, 1985; Hoskins et al. 1985; Boyle and Bosart 1986; Whitaker et al. 1988). It has also been recognized that the ocean-influenced planetary boundary layer plays an important role in the rapid development of coastal cyclones (Petterssen et al. 1962; Gall and Johnson 1971; Tracton 1973; Danard and Ellenton 1980; Sanders and Gyakum 1980; Anthes et al. 1983; Gyakum 1983a,b; Rogers and Bosart 1986). The effect of decreased static stability, through the combination of sensible and latent heat fluxes from the ocean surface and cooling aloft with the approach of an upper-level trough, can be an important factor in enhancing deepening rates (Petterssen et al. 1962; Staley and Gall 1977; Sanders and Gyakum 1980; Bosart 1981; Gyakum 1983a,b; Rogers and Bosart 1986; Atlas 1987; Uccellini et al. 1987). The latent heat release associated with precipitation and fueled by the moisture and heat fluxes in the oceanic planetary boundary layer also contribute to rapid cyclogenesis (Danard 1964; Johnson and Downey 1976; MacDonald and Reiter 1988), especially near the East Coast of the United States (Uccellini et al. 1987). Whether the onset of convection plays a role in triggering rapid cyclogenesis, as hypothesized by Tracton (1973), Bosart (1981), Gyakum (1983a,b), and Anthes et al. (1983), remains an issue that is presently unresolved. Reduced frictional drag over the ocean as compared with land surfaces (Anthes and Keyser 1979; Danard and Ellenton 1980; Rogers and Bosart 1986) may also enhance cyclogenesis and its associated pressure falls at the surface, but this remains uncertain.

### 4.2.4   Sea-Level Pressure Gradient as a Measure of Intensity

The sea-level pressure gradient surrounding a low-pressure center serves as an additional measure of the intensity. The maximum gradients and their directions relative to the centers were examined over a 24-hour period that corresponds to the 24- to 48-hour interval shown for the central sea-level pressures in Fig. 9.[11] Information on the maximum sea level gradients during this time frame, over an area extending outward 500 km from the center of the low, are presented in Table 5. This includes the percentage change in maximum sea-level pressure gradient during the 24-hour examination period for each case, a ranking of the maximum gradients with respect to the total sample of 20 storms, and the direction of the maximum gradient relative to the cyclone center.

The large increases in maximum sea-level pressure gradients (Table 5) are consistent with the tendency of these systems to deepen rapidly. Rapidly falling pressures at the centers of these storms, the decreasing distance between the propagating and developing cyclones, and the persistent nearby anticyclones, whose high pressures are slow to erode over colder land surfaces, result in increased pressure gradients, especially north and east of the developing low-pressure centers. During the 24-hour period examined, the maximum gradients surrounding the low centers increased by 40–500 percent (Table 5), with the gradients nearly tripling on average. The most intense gradients were exhibited by the storms of 10–13 December 1960, 5–7 February 1978,

---

[11] This period was selected because (1) the sea level cyclones typically underwent a significant portion of their total deepening during this time (see Fig. 9) and (2) the low centers were located along the Middle Atlantic or New England coasts, producing heavy snowfall in at least one of the major urban centers between Washington, D.C., and Boston, Massachusetts (see the third, fourth, and fifth panels of the surface analyses shown throughout Chapters 7 and 8).

and 8–10 February 1969, while the weakest gradients occurred in the cases of 22–28 February 1969, 19–21 January 1978, and 18–20 March 1956. The majority of storms developed the largest gradients in their direction of motion, northeast of the low-pressure center. Strong northeasterly winds were usually observed in this quadrant.

As these storms propagated northeastward along the coast and intensified, winds increased to gale force, or in some instances hurricane force, along the immediate coastline. Although wind speed and direction are influenced by friction, isallobaric effects, and the flow of air relative to a moving storm system, the large increases in pressure gradient associated with rapid deepening reliably indicate the high wind speeds that often accompany heavy snowfall.

## 4.3 COLD AIR NEAR THE EARTH'S SURFACE AND ITS SOURCES

In the previous section, cyclogenesis was demonstrated to be an integral component of the 20 major snowfalls. In this section, the surface *anticyclones* that provided and maintained cold temperatures for the heavy snow events are described together with a simple measure of surface temperatures that accompanied these storms.

### 4.3.1 Temperatures during Snowfall Events

Heavy snowfall along the Northeast coast occurs within a wide range of temperatures at the Earth's surface. At temperatures near freezing, snow clings to many surfaces, presenting a hazard for wires and shrubbery that are unable to withstand the load from heavy snow accumulations. At temperatures of about −5°C or lower, snow tends to be powdery and drifts easily in the strong winds that often accompany these storms, creating dangerous traveling conditions.

A representative mean surface temperature was calculated for each storm (Table 6). The mean temperatures were computed from hourly temperature observations during the period when snow was falling at the five major cities[12] that span the Northeast region. Only those cities that reported snowfall exceeding 15 cm were included. Thus, this measure is weighted toward surface temperature conditions in the urban centers.

For the 20-case sample, temperatures accompanying the snowfalls ranged generally between −10°C and 0°C. The coldest storms, as viewed in Table 6, include the cases of January 1964, January 1966, February 1967, and February 1979, while the representative temperatures averaged below −7°C. The warmest storms occurred in March 1958 and late February 1969, when snow fell at temperatures near or slightly above 0°C.

### 4.3.2 Surface Anticyclones

To identify consistent patterns in anticyclone location that characterize the 20-case sample, high-pressure centers were plotted in Fig. 11 for a 36-hour period immediately

---

[12] These cities include Washington, D.C.–National Airport, Baltimore, Maryland, Philadelphia, Pennsylvania, New York, New York–La Guardia Airport, and Boston, Massachusetts.

TABLE 6. Characteristics of cold air outbreaks and anticyclones accompanying each storm, including a measure of mean surface temperatures (°C) for the five major cities during the storm period (see text), the maximum sea-level pressures of anticyclones influencing the Northeast coastal region (Fig. 11) during the 36-hour period described in the text, and the occurrence of damming (indicated by the presence of a wedge in the sea-level pressure field between the Appalachians and the Atlantic coast) and coastal frontogenesis. Occurrences of thunderstorms for at least one of the airport locations in Boston, New York, Philadelphia, Baltimore, or Washington, D.C., are also noted.

| Date | Surface temperature measure (°C) | Maximum sea-level pressures in anticyclones | Cold air damming | Coastal frontogenesis | Occurrence of thunderstorms |
|---|---|---|---|---|---|
| A  18–20 Mar 1956 | −3.1 | 1032 | X | — | X |
| B  14–17 Feb 1958 | −3.8 | 1034 | X | X | X |
| C  18–21 Mar 1958 | +1.5 | 1042 | X | X | X |
| D  2–5 Mar 1960 | −4.6 | 1045 | X | X | X |
| E  10–13 Dec 1960 | −5.7 | 1041 | X | X | — |
| F  18–20 Jan 1961 | −6.8 | 1032 | — | — | — |
| G  2–5 Feb 1961 | −3.2 | 1046 | X | X | — |
| H  11–14 Jan 1964 | −7.5 | 1046 | X | X | — |
| I  29–31 Jan 1966 | −7.3 | 1058 | X | X | — |
| J  23–25 Dec 1966 | −3.7 | 1041 | X | — | X |
| K  5–7 Feb 1967 | −7.6 | 1036 | X | X | X |
| L  8–10 Feb 1969 | −0.6 | 1028 | X | X | — |
| M  22–28 Feb 1969 | −1.5 | 1040 | X | X | — |
| N  25–28 Dec 1969 | −1.4 | 1035 | X | X | — |
| O  18–20 Feb 1972 | −0.1 | 1035 | X | X | X |
| P  19–21 Jan 1978 | −1.7 | 1045 | X | X | — |
| Q  5–7 Feb 1978 | −3.6 | 1053 | X | — | X |
| R  18–20 Feb 1979 | −8.1 | 1050 | X | X | — |
| S  5–7 Apr 1982 | −3.1 | 1033 | X | X | X |
| T  10–12 Feb 1983 | −6.1 | 1040 | X | X | X |

prior to and during cyclogenesis along the East Coast.[13] Maximum pressures were noted in Table 6.

Several patterns emerge from the variety of anticyclone paths prior to and during these intense snowstorms. Anticyclone centers typically followed paths from central Canada, either (i) eastward across Ontario and Quebec, toward the Maritime provinces or New England, or (ii) southward toward the northern and central Plains states. Maximum sea-level pressures within these anticyclones exceeded 1030 mb, and on occasion surpassed 1050 mb (Table 6).

[13] This 36-hour period corresponds to the second through fifth panels of the analyses shown in Chapters 7 and 8.

# PATHS OF ANTICYCLONES AT SEA-LEVEL

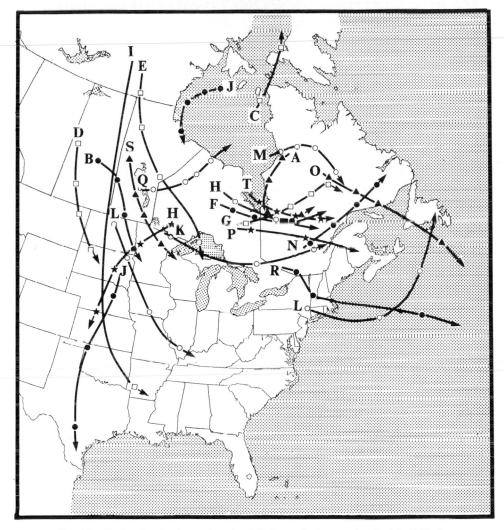

FIG. 11.    Paths of anticyclones at sea level. Letters refer to cases described in Table 6.

The two scenarios outlined above are illustrated by the schematics in Fig. 12. Two basic anticyclone orientations and their associated surface wind streamlines are depicted for heavy snow events along the Northeast coast.[14] In Fig. 12a, the anticyclone

[14] The schematics were derived from patterns of sea level pressure, wind, and temperature associated with the anticyclones and developing cyclones, and analyses of streamlines drawn for selected cases (not shown).

## SEA-LEVEL PRESSURE AND SURFACE STREAMLINES FOR REPRESENTATIVE ANTICYCLONE CONFIGURATIONS DURING MAJOR NORTHEAST SNOWSTORMS

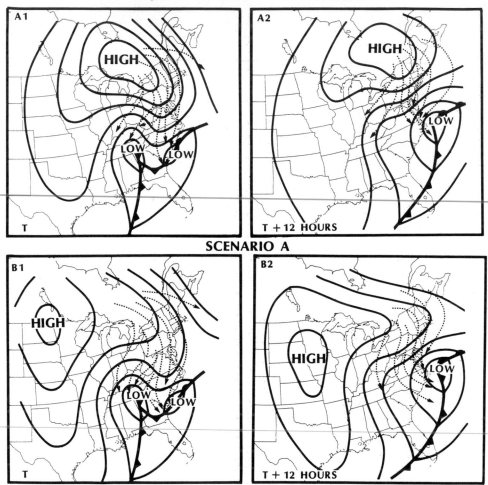

FIG. 12.    Schematic of representative streamlines (dotted arrows) and sea level isobars (solid) for anticyclone–cyclone couplets during major snowstorms along the Northeast coast. Top and bottom panels reflect 12-hour sequences. (A1, A2) shows a well-defined anticyclone over Ontario and Quebec and (B1, B2) shows a high-pressure cell over the Plains states ridging eastward to the northeastern United States.

remains poised to the *north-northwest* of New England while the cyclone develops farther south along the coast.[15] In Fig. 12b, a high-pressure ridge extends across the northeastern United States from an anticyclone located over the Canadian or American

[15] The cases of February 1961, the "Presidents' Day Storm" of February 1979, and the "Megalopolitan Storm" of February 1983 are representative of this scenario. See the corresponding surface analyses in sections 7.7, 7.18, and 7.20.

Plains.[16] These sea-level pressure configurations produce a north-northeasterly flow of air along the coast that drives cold air southward toward the developing cyclone. In both scenarios, the low-level airflow reaching the coast originates in an air mass characterized by temperatures well below seasonal normals. The orientation of the anticyclone relative to the cyclone results in a low-level airflow that passes primarily over land, or for only a very limited distance over the ocean before reaching the coastal sections of the northeastern United States. As a result, the cold air mass is not substantially modified by sensible and latent heating from the nearby ocean. The anticyclones also play a role in cyclogenesis, by maintaining or enhancing the low-level thermal fields and thermal advection patterns along the coastline. These factors contribute to cyclone development and deepening rates.

Despite the frequency of cold air outbreaks across the northeastern United States during the winter season and the passage of cyclonic weather systems about every 3 to 5 days, they only rarely combine to produce heavy snowfall along the Northeast coast. The patterns of sea level pressure, wind, and temperature described here indicate that the position, orientation, and evolution of the anticyclone, relative to the coastline and the developing cyclone, are important factors for maintaining a cold low-level airflow either toward or along the coast.

Heavy snowfall is less likely when the storm center passes nearby or to the north and west of a given location, since the cyclone's counterclockwise circulation draws in warmer air from the south and east. Heavy snow amounts also may not materialize when the anticyclone north of the region continues drifting eastward, producing a broad low-level flow of air passing over the ocean before reaching the coast. In these situations, precipitation often falls as rain along the immediate coast, with heavy snow more likely 100 to 500 km inland. In 2 of the 20 storms examined, December 1969 and February 1972, the combination of a surface low-pressure center very close to the coast and an anticyclone located farther north and east than in the other cases, set up a long easterly fetch of air into the coastal sections of the Northeast. In both cases, precipitation in the coastal regions began and ended as snow, but fell as rain for a significant period in the interim.

### 4.3.3 Cold Air Damming and Coastal Frontogenesis

Cold air damming and coastal frontogenesis are linked to the topography of the eastern United States, and have an important influence on major snow events along the Northeast coast. Cold air at low levels of the atmosphere frequently becomes trapped or "dammed" between the Appalachian Mountains and the Atlantic coast when a cold surface anticyclone passes to the north of the Middle Atlantic states and southern New England (see Baker 1970; Bosart et al. 1972; Richwien 1980, Forbes et al. 1987; Stauffer and Warner 1987; and Bell and Bosart 1988). Damming is evidenced by a narrow wedge of cold air that persists east of the Appalachians in association with significantly ageostrophic flow near the Earth's surface. Damming develops as cold

---

[16] This scenario is found in the storms of March 1960, the January 1966 "Blizzard of '66," and the snowstorm of February 1978. See the corresponding surface analyses in sections 7.4, 7.9, and 7.17.

## EVOLUTION OF COLD AIR DAMMING AND COASTAL FRONTOGENESIS
## PRIOR TO A MAJOR NORTHEAST SNOWSTORM

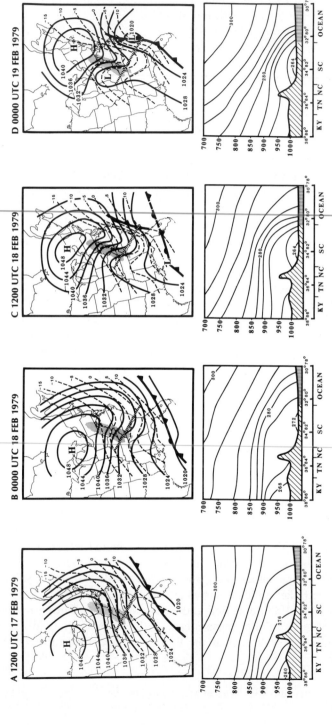

FIG. 13.   Example of a cold air damming event and coastal frontogenesis prior to the February 1979 "Presidents' Day Snowstorm" at (a) 1200 UTC 17 February, (b) 0000 UTC 18 February, (c) 1200 UTC 18 February, and (d) 0000 UTC 19 February 1979. (Top) Surface fronts, sea-level isobars (solid, mb), surface isotherms (dashed, °C), and shading represents location of the Appalachian Mountains. Thick dotted line represents axis of cross sections shown at the bottom. (Bottom) Cross sections of potential temperature (solid, °C).

air supplied by the anticyclone to the north is channeled southward along the eastern slopes of the Appalachians. The cold air wedge is then maintained by a combination of frictional and isallobaric effects, evaporation, adiabatic ascent, and orographic channeling (Forbes et al. 1987; Bell and Bosart 1988).

Damming is represented in the sea-level pressure field by an inverted high-pressure ridge between the mountains and the coast (see left-hand panels in Fig. 12), which may extend as far south as Florida. Its distinctive appearance on the surface weather charts is especially evident in the cases of February 1961, January 1964, February 1979, and February 1983 (see surface analyses in sections 7.7, 7.8, 7.18, and 7.20). Although quantifying the impact of damming is difficult, the inverted sea level pressure ridge and cold air wedge were observed, to some degree, in each of the 20 cases, with the possible exception of January 1961 (Table 6).

Dammed cold air suppresses low-level warm air advection over the coastal plain, thereby maintaining the thermal gradients and warm air advection pattern along the coastline or farther offshore. As a result, when a primary cyclone and its upper-level forcing approach the Appalachians from the west, the low-level warm air advection pattern shifts to the east of the cold air wedge. This process facilitates the formation of a secondary cyclone along the East Coast as the primary system weakens in the Ohio Valley.

As a cold air mass passes across the eastern United States, the eastern flank of its associated anticyclone moves over the Atlantic Ocean. A portion of the airstream connected with the anticyclone often flows from an easterly to northeasterly direction over the western Atlantic, while low-level winds over the coastal plain retain a more northerly component, as a result of cold air damming. The deformation and conver gence of these airstreams result in a zone of enhanced low-level baroclinicity near the coastline, since the air mass over the land remains cold, while air over the ocean is modified by sensible and latent heat fluxes within the ocean-influenced planetary boundary layer. These developments describe the process of "coastal frontogenesis" (Bosart et al. 1972). In general, the coastal fronts that form during damming episodes are easily identified by an inverted sea level pressure trough near, or just offshore of, the coast, and are marked by a distinct temperature contrast and changes in wind direction. These characteristics of coastal frontogenesis were observed in 16 of the 20 cases examined here.

Coastal frontogenesis can play an important role in the development of Northeast snowstorms since it provides a site where low-level convergence, baroclinicity, warm air advections, and surface vorticity are maximized (Bosart 1975). Thus, the coastal front becomes a channel along which the surface cyclone rapidly develops and prop-agates. Notable examples include the storms of February 1961 (especially at 1200 UTC 3 February), January 1964 (especially at 1200 UTC 12 January), February 1979 (especially at 0000 UTC 19 February), and February 1983 (especially at 1200 UTC 10 February). Studies have shown that the convergence associated with coastal front-ogenesis is probably responsible for enhancing the precipitation rate along the cold side of the frontal boundary, where the heaviest snowfall is often observed (Bosart et al. 1972; Bosart 1975; Marks and Austin 1979; Bosart 1981).

A series of sea-level pressure and surface temperature analyses and selected vertical cross sections was drawn for the 36-hour period preceding the development of the February 1979 "Presidents' Day Storm," to illustrate the evolution of damming and coastal frontogenesis along the East Coast prior to a major snowstorm (Fig. 13). Between 1200 UTC 17 February and 0000 UTC 19 February 1979, a large anticyclone accompanied by record- or near-record-setting minimum surface temperatures drifted slowly eastward from the upper Great Lakes to New York. During this period, significant changes in the orientation of the sea level pressure field and surface isotherms indicated damming conditions. By 1200 UTC 18 February 1979, anticyclonically curved isobars and isotherms reflect the highest pressures and lowest temperatures located between the Appalachians and the coast. The coldest temperatures were nearly collocated with the ridge axis, as discussed by Baker (1970) and Richwien (1980). Meanwhile, the thermal gradient along the Southeast coast intensified, as temperatures decreased over the land and increased offshore, indicating the development of a coastal front. Cold air damming and coastal frontogenesis are also depicted by the vertical cross sections of potential temperature in Fig. 13. After 1200 UTC 17 February, decreasing potential temperatures near the Earth's surface created the characteristic "cold dome" [i.e., see Fig. 14 in Bell and Bosart (1988)], as the shallow, cold air mass wedged between the mountains and coastline. The increasing horizontal potential temperature gradient along the coast identified the coastal frontogenesis.

Anticyclones over Canada, cold air damming east of the Appalachian Mountains, and coastal frontogenesis along or near the East Coast have a significant bearing on the development of heavy snowstorms. These features maintain cold air to sustain snowfall and provide an enhanced low-level baroclinic zone that shifts the cyclone path toward a track close to or slightly offshore of the Atlantic coast.

The forecast of damming events by operational large-scale numerical models is still a problem (Richwien 1980; Forbes et al. 1987), possibly because of inadequate vertical resolution of the atmosphere's planetary boundary layer and a poor representation of topography. The inability of the models to properly resolve damming yields a misrepresentation of the lower-tropospheric thermal field, resulting in temperature forecasts that are too high and contributing to the difficulty in delineating frozen versus liquid precipitation along the coastal plain.

## 4.4   OBSERVATIONS OF THUNDERSTORMS DURING HEAVY SNOWSTORMS

Thunderstorms associated with heavy snowfall have been observed in many recent snowstorms along the East Coast of the United States (Kocin et al. 1985; Bosart and Sanders 1986). A cursory examination of surface reports from airport locations in Boston, New York, Philadelphia, Baltimore, and Washington indicates that lightning and thunder were noted in at least one city during 10 of the 20 cases in this study (Table 6). Vivid displays of lightning were widely reported during the Christmas Eve snowstorm of December 1966 and in the February 1983 "Megalopolitan Storm."

Thus, it appears that electrical activity is relatively common in these severe snowstorms, although this conclusion is based on surface reports from only five sites. Bosart and Sanders (1986) have linked the evolution of thunderstorms during the February 1983 snowstorm to rapidly moving mesoscale wave phenomena with gravity wave characteristics. It is not known whether gravity waves were associated with the other cases in which electrical phenomena were observed.

# CHAPTER 5

# Description of Upper-Level Features Associated with the Snowstorms

Upper-level features associated with the 20 snowstorms are examined in this chapter. The analyses are derived from 850-, 700-, 500-, 300-, and 200-mb NMC charts. They are designed to provide an overview of the geopotential height, wind, and temperature fields that are indicative of the upper-level dynamical and thermodynamical processes associated with the storms.

The study of upper-level conditions required for surface cyclogenesis dominated meteorological research from the 1930s through the 1950s, concurrent with the introduction of techniques to monitor and measure atmospheric winds, temperature, and moisture above the Earth's surface. These measurements, combined with concepts promoted by Sutcliffe (1939, 1947), Bjerknes and Holmboe (1944), Charney (1947), Sutcliffe and Forsdyke (1950), Bjerknes (1951), Palmen (1951), Petterssen (1955, 1956), and others, permitted weather forecasters to identify distinct upper-atmospheric features and patterns that contribute to the development of cyclones.

The development of cyclonic disturbances at sea level requires upper-level divergence to achieve a net reduction of mass, thereby reducing sea level pressures in a region where the low-level wind field is generally convergent. The net difference between the divergence aloft and the low-level convergence is a measure of the rate of surface cyclonic or anticyclonic development, as deduced by Dines (1925), emphasized by Sutcliffe (1939, pp. 518–519), and later used to formulate a "development equation" (Sutcliffe 1947) that was further refined by Petterssen (1956). Bjerknes and Holmboe (1944) related the vertical distribution of divergence necessary to sustain cyclogenesis to the presence of upper-level trough/ridge patterns, with divergence aloft, convergence near the Earth's surface, and a level of nondivergence in the middle troposphere downstream of the wave trough axis (Fig. 14a). A reversal of this pattern contributes to anticyclogenesis upstream of the trough axis.

The recognition that upper-level divergence is necessary for surface cyclogenesis was coupled with the requirement for a significant ageostrophic wind component (Sutcliffe 1939; Bjerknes and Holmboe 1944). Bjerknes and Holmboe related the pattern of upper-level divergence associated with upper-level trough/ridge systems in the geopotential height field to "longitudinal" or along-stream ageostrophic wind components resulting from cyclonic and anticyclonic curvature. Subgeostrophic flow at

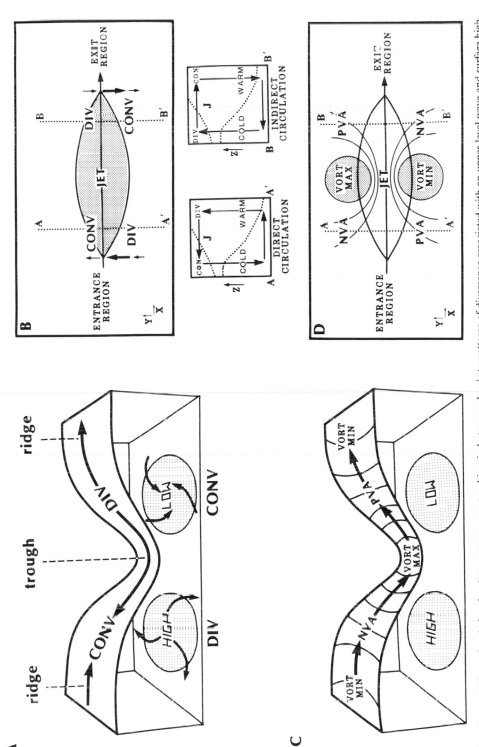

Fig. 14. (a) Schematic relating the along-stream ageostrophic wind at upper levels to patterns of divergence associated with an upper-level wave and surface high-and low-pressure couplets [after Bjerknes and Holmboe (1944)]. (b) Schematic of transverse ageostrophic wind components and patterns of divergence associated with the entrance and exit regions of a straight jet streak [after Bjerknes (1951)]. Vertical cross sections illustrate vertical motions and direct and indirect circulations in the entrance region (line A-A') and exit region (line B-B') of a jet streak. Cross sections include two representative isentropes (dotted), upper-level jet location (J), relative positions of cold and warm air, upper-level divergence, and horizontal ageostrophic wind components within the plane of each cross section. (c) Schematic of maximum (cyclonic) and minimum (anticyclonic) relative vorticity centers and advections (nva = negative or anticyclonic vorticity advection; pva = positive or cyclonic vorticity advection) associated with an idealized upper-level wave. (d) Schematic of maximum and minimum relative vorticity centers and associated advection patterns associated with a straight jet streak.

the base of a trough and supergeostrophic flow at the crest of a ridge combine to enhance the divergence (convergence) downstream (upstream) of the trough axis (Fig. 14a). Bjerknes (1951) also discussed the likely contribution to surface cyclonic and anticyclonic development from the "transverse" or cross-stream ageostrophic wind components associated with upper-level wind maxima known as jet streaks [see Palmén and Newton (1969, p. 199)], since they enhance and focus upper-level divergence and the resultant vertical motion fields (Fig. 14b) (Bjerknes 1951; Riehl et al. 1952; and others).

Scientists recognized that measuring the divergence profile associated with cyclogenesis was impractical [e.g., see Austin (1951, p. 632) and Petterssen (1956, p. 294)], since upper-level wind measurements at the time these concepts were formulated were not sufficiently accurate to diagnose the relatively small difference between the upper- and lower-level divergence that governs cyclogenesis. Since divergence was difficult to measure due to unreliable and scarce wind measurements, kinematic and dynamical diagnostics were used to approximate the divergence by applying the vorticity and thermodynamic equations. These approaches related sea level cyclone development to cyclonic vorticity advection in the middle and upper troposphere [see Sutcliffe (1939, 1947) and Holton (1979, pp. 119–143)], which could easily be diagnosed from the geopotential height fields. Using this method, the divergence can be inferred from patterns of vorticity advections in upper troughs and ridges (Palmén and Newton 1969, pp. 316–318), where maxima and minima in the vorticity fields are typically found, as illustrated in Fig. 14c. The contribution of jet streaks to the divergence could also be inferred from the vorticity-advection patterns associated with the enhanced wind shears on the cyclonic and anticyclonic sides of the jet (Fig. 14d).

Charney (1947) recognized the importance of Bjerknes and Holmboe's work relating upper-level troughs to surface cyclogenesis, but also noted that many such features are observed without surface cyclogenesis. Charney (1947) and Eady (1949) introduced the concept of baroclinic instability, which identified preferred wavelengths that were likely to produce cyclonic development and distinguished between troughs which induce surface cyclogenesis from those that do not. The baroclinic instability theory also points to factors other than upper-level troughs that help to explain why, at certain wavelengths, the upper-level trough/ridge systems amplify, with corresponding surface cyclonic development. The thermal structure and related advection patterns (cold air advection upstream of a trough and warm air advection downstream of the trough) are noted as important elements in the energy transformations (eddy potential to eddy kinetic) required to produce the amplifying wave systems normally associated with surface cyclogenesis.

Based on the development equation, Sutcliffe and Forsdyke (1950) and Petterssen (1955, 1956) offered the rather simple concept that cyclone formation is favored when an upper-level trough, marked by a cyclonic vorticity maximum and cyclonic vorticity advections, approaches a frontal system or low-level baroclinic zone, where thermal advections are large. Sutcliffe and Forsdyke hypothesized that the low-level thermal advection pattern could act to enhance the upper-level vorticity advections (and thus the divergence) in a manner which they termed "self development" [also discussed by Palmén and Newton (1969, pp. 324–326) and reviewed in section 6.1.6]. The vorticity and thermal advection concepts have since been combined within the quasi-geostrophic framework [for example, see Holton (1979, Chapter 7)] and have been related to the horizontal structure of tropospheric waves and vertical circulation patterns that constitute cyclogenesis. An outgrowth of these efforts was the recognition that a baroclinic environment marked by upper-level troughs, large thermal gradients, high wind speeds associated with the jet stream and embedded jet streaks, and associated vertical and horizontal wind shears, is essential for the initiation and intensification of cyclones at the Earth's surface. Since each of the 20 snowstorms was associated with significant cyclonic development, the following discussion of the structure and evolution of their upper-level features will focus on characteristics of the upper-level troughs, winds, vorticity, and temperature.

The 500-mb geopotential height and vorticity fields and the 200- and 300-mb wind fields are presented to describe the trough/ridge systems and jet streaks associated with each storm. The complex structure of the geopotential height and wind fields defies a simple description as noted by Palmén and Newton (1969, p. 273) in their review of numerous cyclone studies; nevertheless, several generalizations will be offered for the 20 cases. Finally, a description of the 850-mb temperature, wind, and moisture fields will illustrate the lower-tropospheric thermal and moisture advections during cyclogenesis.

## 5.1   500-MB TROUGHS

An examination of the 500-mb surface shows that *three* patterns of geopotential height define the upper-level troughs associated with the sea-level cyclones in this 20-case sample (see Table 7). These patterns include (i) the transformation of an "open-wave" trough, in which there are no closed contours at 500 mb, into a closed circulation or vortex during cyclogenesis at sea level (12 of 20 cases); (ii) the existence of a closed circulation at 500 mb prior to and during cyclogenesis (4 of 20 cases); and (iii) 500-mb troughs that did not evolve into closed circulations (4 of 20 cases).

The majority of cases involved the transformation of an open-wave trough into a closed circulation at 500 mb, a process depicted schematically by Palmén and Newton (1969, p. 326). This scenario is demonstrated by the evolution of the 500-mb geopotential height field during the 5–7 April 1982 snowstorm (Fig. 15a), and reflects the "self-development" processes (Sutcliffe and Forsdyke 1950) illustrated by Palmén and Newton (1969, pp. 324–326). The self-development concept describes how the temperature advections and latent heat release associated with cyclogenesis enhance the amplitude of upper-level waves and decrease the wavelength between the trough and ridge axes. The increased amplitude of the upper-level waves and their decreased wavelength produce greater upper-level divergence, which further enhances surface cyclogenesis. The 500-mb vortex forms as the rapidly developing sea-level cyclone appears to grow upward while it occludes. This sequence of events was described by Petterssen (1956, p. 244) and by Johnson and Downey (1976), utilizing an analysis of angular momentum. Following the formation of the upper low, the sea-level and upper-level circulation centers become nearly vertically aligned as the temperature gradient decreases, marking the decaying stage of the cyclone's evolution.

Among the 20 storms studied, four cases featured a closed circulation in the

TABLE 7. **Characteristics of the 500-mb geopotential height field, including a classification of trough evolution into either open-wave (O) or closed-contour (C) structure, measures of the amplitude and half-wavelength changes (percentage increase or decrease in distance) between the trough and downstream ridge (see text for details), the presence or development of diffluence downwind of the trough axis and a "negatively tilted" trough axis (oriented from northwest to southeast), and the existence of multiple centers of absolute vorticity that are associated with the 500-mb troughs.**

| Date | Trough evolution | Measure of amplitude change (%) | Measure of half-wavelength change (%) | Diffluence | Negative tilt | Multiple vorticity centers |
|---|---|---|---|---|---|---|
| 18–20 Mar 1956 | O → C | +200 | −22 | X | X | |
| 14–17 Feb 1958 | O → C | +44 | −64 | X | X | X |
| 18–21 Mar 1958 | O → C | +121 | −52 | X | X | X |
| 2–5 Mar 1960 | O → C | +7 | −46 | X | X | X |
| 10–13 Dec 1960 | C → O → C | +24 | −35 | X | X | |
| 18–20 Jan 1961 | O → C | +368 | −57 | X | X | |
| 2–5 Feb 1961 | C | +64 | −27 | X | X | |
| 11–14 Jan 1964 | C | +41 | −11 | X | — | X |
| 29–31 Jan 1966 | C | +342 | −21 | X | X | X |
| 23–25 Dec 1966 | O → C | +218 | −14 | X | X | |
| 5–7 Feb 1967 | O | +128 | −2 | X | X | |
| 8–10 Feb 1969 | O → C | +198 | −33 | X | X | |
| 22–28 Feb 1969 | O → C | −46 | −38 | X | X | X |
| 25–28 Dec 1969 | O → C | +290 | −12 | X | X | |
| 18–20 Feb 1972 | O → C | +190 | −17 | X | X | X |
| 19–21 Jan 1978 | O | −12 | −53 | X | X | |
| 5–7 Feb 1978 | C | +420 | −31 | X | X | |
| 18–20 Feb 1979 | O | +31 | −34 | X | X | X |
| 5–7 Apr 1982 | O → C | +233 | −24 | X | X | |
| 10–12 Feb 1983 | O | +30 | −17 | X | X | X |

middle and upper troposphere *prior to* sea-level cyclogenesis (Table 7). Closed upper-level circulations, or "cutoff" lows, may form prior to, simultaneously with, or without surface cyclogenesis, and tend to drift slowly, as pointed out by Petterssen (1956, pp. 244–245) and Palmén and Newton (1969, pp. 274–283). The cutoff low reflects a pool of cold tropospheric air that is detached from the primary region of cold air located at higher latitudes. While a cold pool may sometimes be the remnant of a previously occluded storm, the four cases in this sample were not associated with earlier major surface cyclones. This type of situation is represented by the 500-mb geopotential height analyses during the 12–14 January 1964 snowstorm (Fig. 15b). In this case, a primary and a secondary surface low developed in association with a closed cyclonic circulation aloft as it moved slowly eastward from the central United States to the East Coast.

Although the formation of a cutoff low prior to or during surface cyclogenesis is a prevalent feature of the 20-case sample, 4 storms occurred in connection with a 500-mb trough that did not develop a vortex while the sea-level cyclone was affecting the Northeast coast (see Table 7). The dates of the four cases that exhibited this upper-

# REPRESENTATIVE EXAMPLES OF TROUGH EVOLUTION AT 500 MB FOR MAJOR NORTHEAST SNOWSTORMS

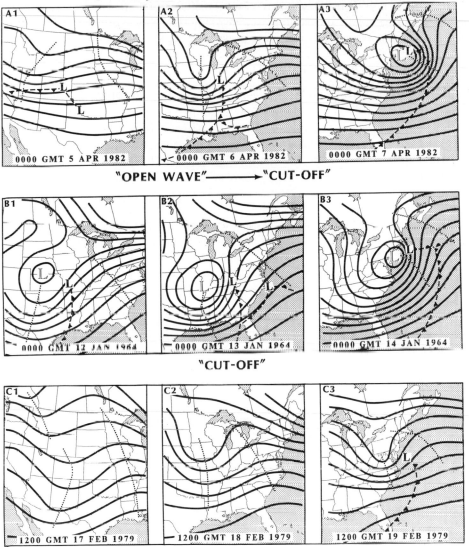

FIG. 15.   Representative examples of trough evolution at 500 mb for major Northeast snowstorms. (A1, A2, A3) Illustration of "open wave" to "cutoff" system, by storm of 5–7 April 1982. (B1, B2, B3) Illustration of cutoff system prior to and during cyclogenesis by storm of 12–14 January 1964. (C1, C2, C3) Illustration of an open wave trough that did not evolve into a cutoff system by storm of 17–19 February 1979. Solid lines are geopotential height contours at 500 mb. Dotted lines indicate trough and ridge axes. Positions of sea-level cyclone and surface fronts are also shown

level pattern are 5–7 February 1967, 19–20 January 1978, 18–19 February 1979, and 10–12 February 1983. Analyses of the 500-mb geopotential height field for the 17–19 February 1979 "Presidents' Day Storm" (Fig. 15c) illustrate that an open-wave system can accompany a rapidly developing cyclone along the coast, without the formation of a closed circulation in the middle and upper troposphere.

### 5.1.1   Amplitude and Wavelength

The amplitude and wavelength of troughs and ridges aloft contribute significantly to upper-level divergence, a well-known requirement for sea-level development. The increased amplitude of upper-level trough/ridge systems enhances vorticity gradients due to changes in curvature, and strengthens vorticity advection along the upper-level flow, which implies an increase of upper-level divergence (Palmén and Newton 1969, p. 325). Increasing amplitude is also indicative of wave growth, an essential component of baroclinic instability theories which describe the energy conversions that take place during cyclogenesis (Charney 1947; Eady 1949; Gall 1976; Farrell 1985). A decrease in the distance, or half-wavelength, between a trough and the downstream ridge can also increase upper-level divergence. Changes in the amplitude and wavelength of upper-level trough/ridge systems were studied to determine whether developing cyclones act as "positive feedback" mechanisms for sustaining or even enhancing cyclogenesis, as described by Sutcliffe and Forsdyke (1950) and Palmén and Newton (1969, pp. 324–326).

The amplitudes of the upper-level trough/ridge systems that spawned these cyclones were examined over the 36-hour period corresponding to the second through fifth panels of the analyses shown in Chapter 7 to capture cyclogenesis along the East Coast in every case.[17] The amplitude is defined as half the mean displacement between the base of the trough and the crest of the downstream ridge. The displacement was computed by selecting three representative 500-mb geopotential height contours that pass close to the vorticity maximum or maxima shown in Fig. 17.

In 18 of the 20 cases (Table 7), the amplitude of the trough/ridge system increased during the 36-hour period. Increases of greater than 100% were observed in 11 cases, with the greatest amplifications shown in the storms of January 1966 (342%), January 1961 (368%), and February 1978 (420%). The large increases in amplitude are not surprising in view of the rapid cyclogenesis that accompanied most of these storms.

The only cases associated with decreasing amplitude were the 19–21 January 1978 and 22–28 February 1969 storms. The January 1978 cyclone did not deepen appreciably while it affected the Northeast (Fig. 9b2, Table 4), and was associated with a trough/ridge configuration of large amplitude *prior* to reaching the East Coast (see 500-mb analyses in section 7.16). The storm of late February 1969 developed as a series of troughs alternately increased and decreased in amplitude prior to the establishment of a large stationary vortex off the New England coast (see section 7.13). In

---

[17] Measures of amplitude and half-wavelength in Table 8 *should not* be considered irrefutable since geopotential height analyses over the ocean are based on limited data. The trough and ridge axes may also be ill defined due to gently curving flow or because of the superpositioning of a shorter-wavelength feature and a longer-wavelength trough or ridge.

the February 1979 "Presidents' Day Storm," the 500-mb trough/ridge system amplified slightly during the 36-hour period encompassing the development of the sea-level cyclone (Table 7). On 19 February, however, the 12-hour period of most rapid intensification was accompanied by a slight decrease in amplitude (see section 7.18). This is a striking deviation from many other cases and conventional views of cyclogenesis.

Changes in the half-wavelength between the troughs and downstream ridge crests are also listed in Table 7. The half-wavelength is defined as the mean distance between the trough axis, as indicated by the maximum cyclonic curvature, and the downstream ridge axis, as indicated by the maximum anticyclonic curvature. The mean distance was derived from measurements along the same three geopotential height contours used to calculate the amplitude changes. The same 36-hour time interval was examined.

In *all* 20 cases, the half-wavelength decreased during cyclogenesis (Table 7), with 36-hour decreases ranging from 2% (February 1967) to 64% (February 1958). Despite concerns about the accuracy of the analyses over the oceans, the consistent depiction of well-defined upper troughs and ridges with shortening wavelengths provides a clear upper-level signature of these storms. The general observation that amplitudes were increasing during the period of shortening wavelength indicates that increases in vorticity advection and upper-level divergence are also characteristic of the cyclogenetic period along the coast. The three cases shown in Fig. 15 are representative of the changes in amplitude and half-wavelength found in the sample.

### 5.1.2   Diffluence and Trough Axis Tilt

The development of diffluence downwind of the trough and a change in the orientation of the trough axis from a positive (northeast–southwest) or neutral (north–south) tilt to a negative tilt (northwest–southeast) are distinguishing characteristics of the upper-level trough evolution in 19 of the 20 cases (Table 7). Diffluence, the lateral divergence of the upper-level height contours, has been recognized as the signature of increasing upper-level divergence immediately downstream of a trough axis (Scherhag 1937; Bjerknes 1954; Palmén and Newton 1969, pp. 334–335). Diffluence becomes more pronounced as (i) the trough amplitude increases and the wavelength decreases, (ii) the trough axis becomes negatively tilted, and (iii) an upper-level jet streak propagates from the west–northwest of the trough axis to the base of the trough. Bjerknes (1951) has also noted this process as an important factor contributing to surface cyclogenesis, as depicted schematically in Fig. 16. The development of a diffluent, negatively tilted trough in the April 1982 storm is depicted by the 500-mb geopotential height analyses in Fig. 15a. In this example, a 500-mb trough progressed from the western United States to the Northeast coast during the 2-day period from 0000 UTC 5 April to 0000 UTC 7 April 1982. Note that the orientation of the trough axis (dotted lines in Fig. 15) changed from northeast–southwest (positive tilt, 5 April), to north–south (neutral tilt, 6 April), to northwest–southeast (negative tilt, 7 April). The evolution of the trough axis from a positive tilt to a negative tilt occurred concurrent with an increase in the trough/ridge amplitude and a decrease in the wavelength, as well as the formation of an intense 500-mb vortex and development at sea level (see surface analyses in section 7.19).

# EVOLUTION OF A DIFFLUENT, NEGATIVELY-TILTED TROUGH DURING CYCLOGENESIS

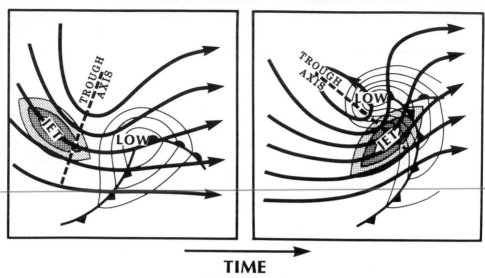

**TIME**

FIG. 16.   Schematic representation of the evolution of a diffluent, negatively tilted upper-level trough during cyclogenesis. Arrows represent upper-level streamlines, dashed line depicts trough axis, and shaded region indicates upper-level jet streak. Thin lines are surface isobars, with surface fronts also shown.

### 5.1.3   Vorticity

The increasing upper-level divergence and vorticity advection that accompany upper-level trough/ridge systems whose amplitudes (wavelengths) are increasing (decreasing) are associated with distinct 500-mb vorticity maxima. The paths of these maxima were plotted for each case in Fig. 17. Locations of maximum absolute vorticity over the 60-hour period covering each storm were obtained from computer-derived analyses generated by NMC and available from NCDC for events since May 1961. For cases prior to May 1961, the positions of the vorticity maxima were derived from analyses performed on a 2° grid using a Barnes (1964) objective analysis scheme and modified for interactive analysis purposes by Koch et al. (1983).

    The vorticity maxima followed a variety of paths, but were initially directed from northwest to southeast in 14 of the 20 cases (see Fig. 17). These equatorward-propagating features indicate deepening or "digging" troughs, in which the amplitude measured between the trough and *upstream* ridge axes was increasing with time. An intensifying ridge is usually present upstream of a digging trough, as noted by Palmén and Newton (1969, p. 335), and is best illustrated by the cases of January 1961, February 1978, and February 1983 (see 500-mb analyses in sections 7.6, 7.17, and 7.20). The deepening troughs were marked by growing cyclonic vorticity due to increasing curvature and cyclonic wind shear associated with the propagation of an upper-level jet from the ridge toward the trough axis (Bjerknes 1951; Palmén and Newton 1969, p. 341).

**PATHS OF ABSOLUTE VORTICITY MAXIMA AT 500 MB**

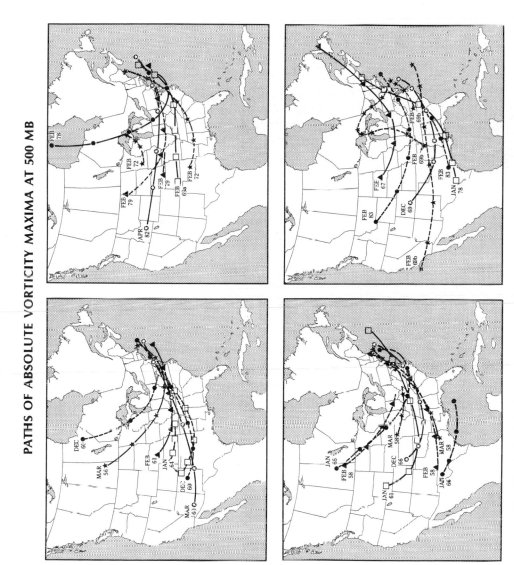

FIG. 17. Paths of absolute vorticity maxima at 500 mb. The 20 cases are grouped as in Fig. 7.

Following the initial deepening or digging of the upper-level troughs in the central United States, the paths of the vorticity maxima typically curved to the east or northeast as they approached the Atlantic coast. This scenario is similar to those described in a climatological study of heavy snowfall over the eastern United States (Goree and Younkin 1966) and in a composite study of rapidly intensifying cyclones over the west-central North Atlantic Ocean (Sanders 1986a). With few exceptions, the centers of maximum cyclonic vorticity moved off the coast between North Carolina and New Jersey to positions just south of the southern New England coast.

Observations spanning several winters indicate that heavy snow seldom occurs over the coastal plain when the vorticity maximum associated with an amplifying upper-level trough curves to the northeast *before* reaching the East Coast. In cases where the vorticity maximum propagates northeastward into the Ohio Valley, the primary surface low-pressure system usually maintains a well-defined circulation west of the Appalachians, advecting lower-tropospheric warm air over the coastal plain. Secondary cyclogenesis still occurs in some cases and cold air may remain in the lowest levels of the atmosphere east of the Appalachian Mountains due to cold air damming. The warm air advection aloft, however, often produces precipitation in the form of rain, freezing rain, or ice pellets, rather than heavy snow, in the major metropolitan areas along the coast. Therefore, the predicted path of the vorticity maximum is an important consideration for forecasters to distinguish heavy snow events from other cases in which substantial snowfall is less likely.

Although the foregoing scenarios suggest a rather straightforward evolution of the deepening and movement of upper-level troughs and their associated vorticity maxima, there are complex variations. Nine of the 20 cases (Fig. 17, Table 7) involved the superpositioning of multiple vorticity centers that underwent various forms of amplification, consolidation, or both (i.e., see Lai and Bosart 1988). The merger of multiple vorticity centers indicates a "phasing" of separate short-wave troughs or jet-stream systems upwind of and prior to sea-level development along the coast. The interaction of these systems enhances the upper-level vorticity advection and associated upper-level divergence as the merging of separate vorticity maxima into one cohesive maximum produces a well-organized pattern of vorticity advection above the developing surface cyclone. Forecasters have long recognized the dilemma in determining whether separate troughs and their associated vorticity maxima will phase or remain separate, especially since numerical prediction models have not always been successful in simulating these interactions (see the discussion for the 22–23 February 1987 snowstorm in section 8.3).

## 5.2   UPPER-LEVEL HEIGHT AND WIND FIELDS ASSOCIATED WITH COLD SURFACE ANTICYCLONES

The 500-mb surface was also examined to identify characteristic geopotential height patterns associated with the sea-level anticyclones over the Great Lakes and southeastern Canada (see Figs. 11 and 12) that provide the cold dry air which is advected southward toward the developing storms.

The sea-level high-pressure systems were found upwind of upper-level troughs propagating across eastern Canada in all 20 cases (Table 8). The paths of their associated vorticity maxima trace the progression of these upper-level troughs (Fig. 18) during the 36 hour period prior to cyclogenesis.[18] Figure 18 shows that the troughs generally crossed eastern Ontario, Quebec, and the Maritime provinces to the offshore waters of the Atlantic Ocean before and during the onset of heavy snowfall in the Northeast coastal region.

A separate circulation feature termed the "Greenland Block," a cutoff upper-level anticyclone centered near Greenland, is often cited by forecasters as a premonitory sign of a flow regime that maintains cold weather in the Northeast and is conducive to snowstorms along the coast. An upper-level anticyclone center, however, was located between Newfoundland and Iceland in fewer than half the cases (see Table 8), although the February 1958, January 1964, and February 1983 cases demonstrated large cutoff anticyclones prior to the snowstorms (see the 500-mb analyses for these cases in sections 7.2, 7.8, and 7.20). The mechanisms that create "blocking" patterns and their roles in establishing conditions conducive to snowfall are related to complex planetary-scale processes (e.g., see Colucci 1985) that are beyond the scope of this study.

TABLE 8.   Characteristics of the 500-mb geopotential height field associated with the surface anticyclone during the 36-hour period prior to cyclogenesis along the East Coast (corresponding to the first through fourth panels of analyses shown in Chapter 7), including the presence of 1) a trough crossing eastern Canada (Quebec and the Maritime provinces), 2) a cutoff anticyclone located near Greenland ("the Greenland Block") defined by at least one closed contour at 500 mb, and 3) confluence upwind of the trough crossing eastern Canada.

|   | Date | Eastern Canadian trough | "Greenland Block" | Confluence |
|---|------|------------------------|-------------------|------------|
| A | 18–20 Mar 1956 | X | — | X |
| B | 14–17 Feb 1958 | X | X | X |
| C | 18–21 Mar 1958 | X | — | X |
| D | 2–5 Mar 1960 | X | X | X |
| E | 10–13 Dec 1960 | X | — | X |
| F | 18–20 Jan 1961 | X | — | X |
| G | 2–5 Feb 1961 | X | — | X |
| H | 11–14 Jan 1964 | X | X | X |
| I | 29–31 Jan 1966 | X | X | X |
| J | 23–25 Dec 1966 | X | — | X |
| K | 5–7 Feb 1967 | X | — | X |
| L | 8–10 Feb 1969 | X | — | — |
| M | 22–28 Feb 1969 | X | — | X |
| N | 25–28 Dec 1969 | X |   | X |
| O | 18–20 Feb 1972 | X | — | X |
| P | 19–21 Jan 1978 | X | — | X |
| Q | 5–7 Feb 1978 | X | X | X |
| R | 18–20 Feb 1979 | X | X | X |
| S | 5–7 Apr 1982 | X | X | X |
| T | 10–12 Feb 1983 | X | X | X |

[18] This period corresponds to the first through fourth panels presented in Chapters 7 and 8.

## PATHS OF ABSOLUTE VORTICITY MAXIMA AT 500 MB: EASTERN CANADIAN TROUGHS

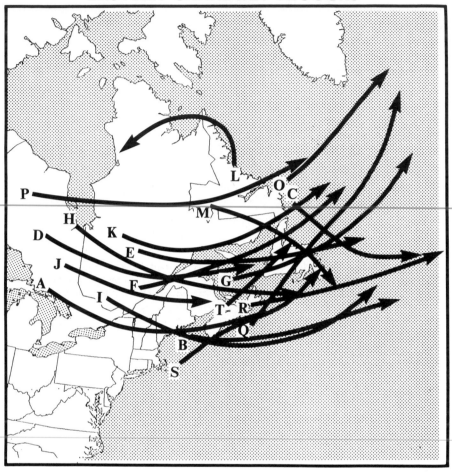

FIG. 18. Paths of absolute vorticity maxima associated with troughs moving across eastern Canada. Letters refer to cases described in Table 8.

In 19 of the 20 cases (Table 8), the sea-level anticyclone was located beneath a region of confluent geopotential heights, as illustrated by the schematic in Fig. 19. Confluence, defined as the lateral convergence of the upper-level height contours, was found upwind of the 500-mb trough over eastern Canada, which typically contained a distinct geopotential height minimum or closed circulation. The association between upper-level confluence and the location of the surface anticyclone is best illustrated by the cases of February 1961, February 1979, and February 1983 (see sections 7.7, 7.18, and 7.20), but is not as evident in such other cases as March 1956 and December 1969 (see sections 7.1 and 7.14). There was no confluence zone in the 8–10 February

## REPRESENTATION OF UPPER-LEVEL CONFLUENCE, JET STREAK AND ASSOCIATED SURFACE ANTICYCLONE

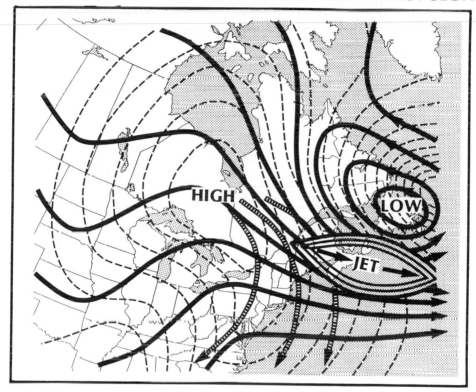

FIG. 19.   Schematic representation of upper-level confluence (illustrated by solid arrows), jet streak (depicted by the thick solid lines and short arrows), and surface anticyclone (sea-level isobars are shown as thin dashed lines). Stippled arrows represent low-level airflow into the northeastern United States.

1969 case, a snowstorm that was not preceded by a cold anticyclone north-northwest of the developing surface low (see the surface and 500-mb charts in section 7.12).

The repeated appearance of upper-level confluence and its connection with the sea-level anticyclones suggest that the development and maintenance of a confluence zone over the northeastern United States or southeastern Canada help to establish and maintain the lower-tropospheric cold temperatures necessary for snowfall along the coast. Petterssen (1956, p. 333) and Reiter (1963, p. 155; 1969, p. 183) described regions of confluence upwind of upper-level troughs as favorable sites for anticyclone development and the southward transport of low-level cold air.

It is well known that wind speeds associated with the jet stream are enhanced in confluent regimes (Namias and Clapp 1949). It is, therefore, not surprising that the confluence observed in most of these cases was linked to the presence of polar jet streaks across the Northeast (Fig. 19 and Table 9) which are independent of the jet streaks associated with the upper-level troughs approaching the East Coast further

TABLE 9. **Descriptions of the middle- to upper-tropospheric wind fields, including the observed increase of wind speeds ($>15$ m s$^{-1}$) over a period of 24 hours or less (and location), and the presence of upper-level wind maxima at the base of the troughs nearing the East Coast and across the northeastern United States or southeastern Canada during a 12-hour period during cyclogenesis along the East Coast (corresponding to the third and fourth panels of the analyses shown in Chapter 7).**

| Date | Observed increases of wind speed $> 15$ m s$^{-1}$ over 24-hour period or less (and location) | Upper-level jet streaks at base of troughs nearing the East Coast during cyclogenesis | Upper-level jet streaks across the northeastern United States or southeastern Canada during cyclogenesis |
|---|---|---|---|
| 18–20 Mar 1956 | — | X | X |
| 14–17 Feb 1958 | X (downstream ridge) | X | X |
| 18–21 Mar 1958 | X (trough axis) | X | X |
| 2–5 Mar 1960 | X (trough axis) | X | X |
| 10–13 Dec 1960 | X (downstream ridge) | X | X |
| 18–20 Jan 1961 | X (trough axis) | X | X |
| 2–5 Feb 1961 | X (trough axis) | X | X |
| 11–14 Jan 1964 | X (downstream ridge) | X | X |
| 29–31 Jan 1966 | X (downstream ridge) | X | X |
| 23–25 Dec 1966 | — | X | X |
| 5–7 Feb 1967 | X (trough axis) | X | X |
| 8–10 Feb 1969 | X (trough axis) | X | — |
| 22–28 Feb 1969 | — | X | X |
| 25–28 Dec 1969 | — | X | X |
| 18–20 Feb 1972 | — | X | ill-defined |
| 19–21 Jan 1978 | X (downstream ridge) | X | X |
| 5–7 Feb 1978 | — | X | — |
| 18–20 Feb 1979 | X (downstream ridge) | X | X |
| 5–7 Apr 1982 | — | X | X |
| 10–12 Feb 1983 | X (trough axis) | X | X |

south. The following description of middle- and upper-tropospheric winds establishes a link between the jet-stream winds in the confluent region above the surface anticyclone and the diffluent region above the developing surface cyclone, and the evolution of snowstorms along the northeast coast of the United States.

## 5.3 MIDDLE- AND UPPER-TROPOSPHERIC WINDS

Jet streams are integral components of the cyclonic weather systems that generate heavy snowfall in the Northeast. Horizontal wind shears and the resultant large values of absolute vorticity are associated with these narrow meandering bands of strong wind. Embedded within the jet stream are distinct isotach maxima, or jet streaks, as defined by Palmén and Newton (1969, p. 199). The cyclonic wind shear associated with jet streaks focuses and enhances the cyclonic vorticity maximum, vorticity advection, and the associated upper-level divergence and vertical-motion fields (Riehl et al. 1952; Cahir 1971; Uccellini and Johnson 1979) that promote sea-level cyclogenesis (Bjerknes 1951; Riehl et al. 1952; Newton 1956; Reiter 1963, 1969).

"Transverse," or cross-stream, ageostrophic components in the entrance and exit regions of upper-level jet streaks contribute to divergence. Namias and Clapp (1949), Bjerknes (1951), Murray and Daniels (1953), Uccellini and Johnson (1979), and others found that the entrance region of an idealized jet streak is marked by a transverse ageostrophic component directed toward the cyclonic-shear side of the jet (Fig. 14b). This component represents the upper branch of a direct transverse circulation that converts available potential energy into kinetic energy for air parcels accelerating into the jet. The direct circulation is marked by rising motion on the anticyclonic or warm side of the jet and by sinking motion on the cyclonic or cold side of the jet (Fig. 14b). In the exit region, the ageostrophic component is directed toward the anticyclonic-shear side of the jet (Fig. 14b), representing the upper branch of an indirect transverse circulation (Fig. 14b) that converts kinetic energy to available potential energy as air parcels decelerate upon exiting the jet. This circulation features rising motion on the cyclonic or cold side of the jet and sinking motion on the anticyclonic or warm side of the jet. These transverse circulation patterns are consistent with the vorticity advection concepts described by Riehl et al. (1952) and illustrated in Fig. 14d.

The influence of curvature effects in masking the contribution of jet streaks to upper-level ageostrophy and divergence, and their associated vertical motion fields is discussed in recent studies by Shapiro and Kennedy (1981), Newton and Trevisan (1984), Uccellini et al. (1984), Keyser and Shapiro (1986), Kocin et al. (1986), and Uccellini and Kocin (1987). These studies indicate that curvature complicates the simple two-dimensional relationships (as shown in Fig. 14b) between the ageostrophic wind field associated with jet streaks and divergence. Diabatic processes, especially those related to latent heat release, can also alter the vertical motions associated with jet-streak circulations (for example, see Cahir 1971 and Uccellini et al. 1987).

Despite the complications introduced by curvature and diabatic processes, the contribution of jet-streak circulations to surface cyclogenesis has been shown in numerous studies. Hovanec and Horn (1975) and Achtor and Horn (1986) illustrated the importance of the left-front exit region of jet streaks in cyclogenesis to the lee of the Rocky Mountains. Mattocks and Bleck (1986) and Sinclair and Ellsberry (1986) emphasized the role of the indirect circulation in the exit region of upper-level jet streaks in the development of cyclones in the lee of the Alps, the Gulf of Genoa, and the northern Pacific Ocean. Uccellini et al. (1985, 1987) showed how vertical circulations associated with upper-level jet streaks contributed to two periods of surface cyclonic development during the "Presidents' Day Storm." Uccellini and Kocin (1987) related the entrance and exit regions of two separate upper-level jet streaks and their associated direct and indirect circulations to the development of heavy snowfall along the northeast coast of the United States. A discussion of how these direct and indirect circulations contribute to heavy snowstorms in the Northeast will follow a description of the complex evolution of the jet streaks associated with the 20 major snowstorms.

### 5.3.1   Horizontal and Vertical Resolution

Principal jet-stream axes were superimposed on the 12-hourly 500-mb geopotential height analyses shown for each of the 20 cases in Chapter 7. These analyses were derived from the operational 500-, 300-, 250- (where applicable), and 200-mb charts. Isotachs were drawn for the maximum wind speeds at either 500, 300, or 200 mb, with the level of maximum wind indicated in parentheses. The upper-level wind analyses indicate that the sample of 20 storms is characterized by multiple- and multilevel jet systems that cannot easily be described due to their complicated horizontal and vertical structure and evolution. Both polar and subtropical jet streaks were present concurrently in many cases. Polar jet streaks are usually found on the 300-mb isobaric chart, but may also be seen on the 200- and 500-mb charts since they are associated with baroclinic zones that extend through a deep layer of the lower troposphere. Subtropical jet streaks are mainly observed on the 200-mb chart, since they are typically associated with baroclinic regions that are confined to the upper troposphere. Wind analyses of the 20 cases in Chapter 7 show that the polar jets attained speeds of 50–70 m s$^{-1}$, while maximum speeds in the subtropical jet streaks approached 80–90 m s$^{-1}$ in some cases. Wind measurements in the upper troposphere may not always be accurate, however, due to the low elevation angle between the observing system and the radiosonde (Newton and Persson 1962). Also, missing upper-level wind reports, which plagued many of the 20 cases, made identifying and analyzing the jet streak systems more difficult.

The complex structure of jet streams and the jet streaks embedded within them is illustrated by the January 1966 and February 1979 cases (Fig. 20). At 1200 UTC 29 January 1966, the 200-, 300-, and 500-mb charts (Fig. 20a–c) reveal multiple jet streaks at various levels of the atmosphere, with the incipient surface cyclone located over the Gulf Coast states (Fig. 20d; also see analyses in section 7.9). At 200 mb (Fig. 20a), a jet streak extended from the western Gulf of Mexico through the southeastern United States with an 85 m s$^{-1}$ wind speed maximum over Georgia. The concentration of maximum wind speeds in the upper troposphere and their relative equatorward location are characteristic of a subtropical jet streak. At 300 mb (Fig. 20b), jet streaks were analyzed over the Ohio Valley and over Florida. The well-defined wind maximum over the Ohio Valley, with velocities exceeding 85 m s$^{-1}$, did not appear in any form on the 200-mb analyses, although the lower extension of this jet system was evident in the 500-mb analysis over the northeastern states (Fig. 20c), indicating that this was a polar jet. The relative wind speed maximum over Florida at 300 mb may be a lower, sloped extension of the jet streak found over Georgia at the 200-mb level. A separate upper-level jet was noted at 200, 300, and 500 mb over the Plains states on the upwind side of the amplifying trough over the central United States.

The second example of jet system structure is from the February 1979 "Presidents' Day Storm." At 1200 UTC 18 February 1979, the 200- and 300-mb charts (Figs. 20e and 20f) show evidence of multiple jet streaks, at a time when heavy snow was developing across the southeastern United States (Fig. 20h). At 200 mb (Fig. 20e), the eastern half of the United States was located beneath an extensive belt of high winds, with an 85–90 m s$^{-1}$ jet streak analyzed over North Carolina. There was also a small

EXAMPLES OF MULTIPLE UPPER-LEVEL JET STRUCTURE FOR TWO CASES

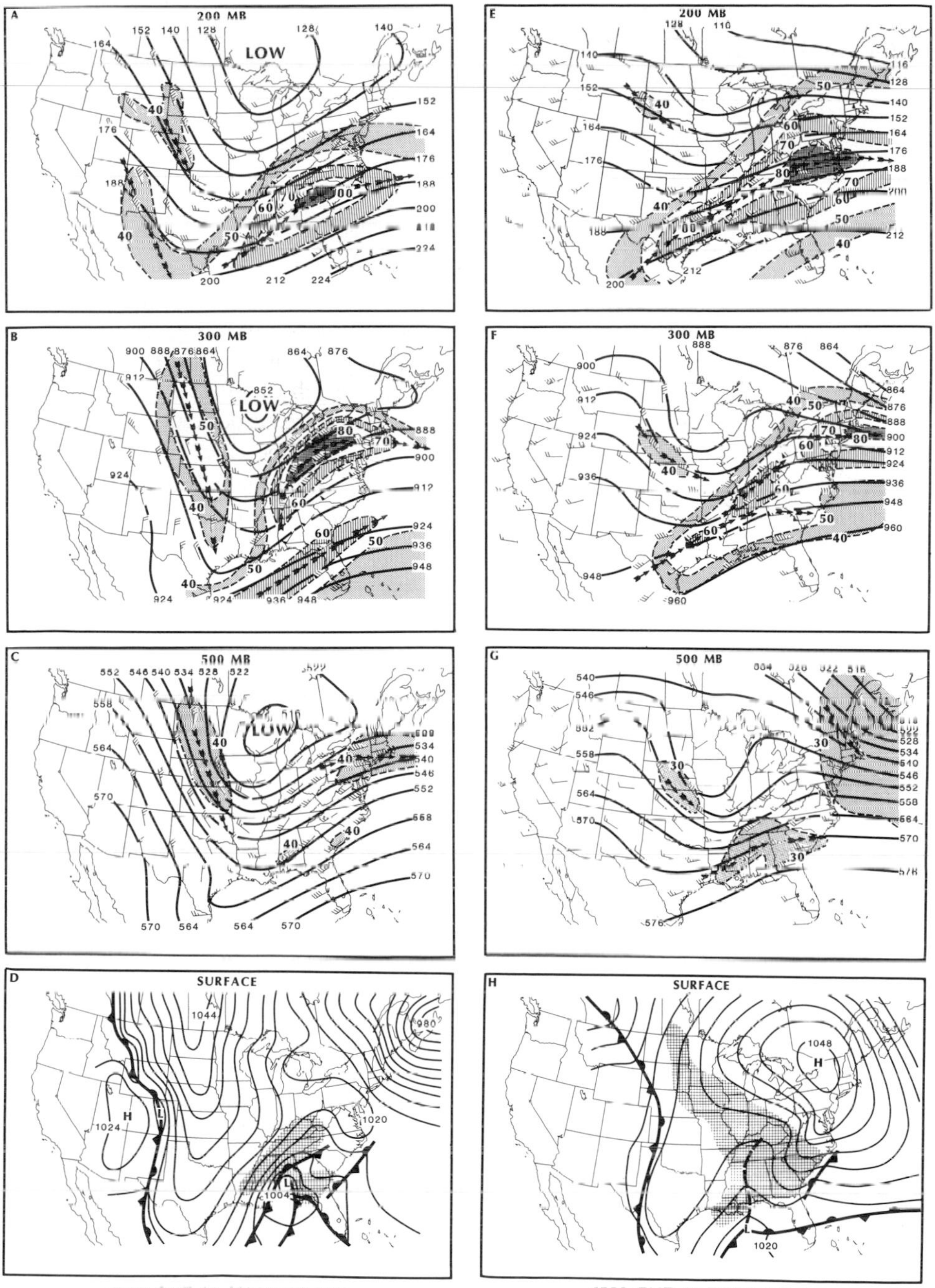

1200 GMT 29 JANUARY 1966          1200 GMT 18 FEBRUARY 1979

FIG. 20.   Examples of multiple upper-level jet structure at (left) 1200 UTC 29 January 1966 and (right) 1200 UTC 18 February 1979. Arrows indicate jet axes. (a, e) 200-mb analyses, including geopotential height (solid, 140 = 1400 m) and isotachs (dashed, m s$^{-1}$, shading for alternating 10 m s$^{-1}$ intervals above 40 m s$^{-1}$). (b, f) 300-mb analyses, including geopotential height (solid, 900 = 9000 m) and isotachs (dashed, m s$^{-1}$, shading for alternating 10 m s$^{-1}$ intervals above 40 m s$^{-1}$). (c, g) 500-mb analyses, including geopotential height (540 = 5400 m) and isotachs (dashed, m s$^{-1}$). (d, h) Surface analyses, depicting fronts, sea-level isobars (thin solid, mb), and precipitation areas.

separate 40 m s$^{-1}$ jet streak over western South Dakota. At 300 mb (Fig. 20f), two jet streaks were evident within the belt of high wind speeds over the eastern United States. An 80 m s$^{-1}$ jet streak was centered over southern New England, while a 65 m s$^{-1}$ streak was located over the Tennessee Valley. The wind speed maximum over New England, which was not readily apparent at 200 mb, represents a descending polar jet within the confluence zone upstream of a trough off the Canadian coast. The speed maximum over Tennessee was actually the lower extension of the ascending subtropical jet streak that had its maximum velocities at 200 mb over North Carolina. The polar jet streak over the Plains states was better defined at 300 mb than at 200 mb, and was also observed at 500 mb (Fig. 20g). For more details of the jet structures in this case, see the analyses in section 7.18 and Uccellini et al. (1984, 1985).

These examples show not only the existence of multiple jet features at a given time for particular cases, but also that the use of only one isobaric chart is insufficient to identify and resolve vertical structure of the jet streaks associated with these snowstorms. The complexity of the upper-level wind patterns and possible interactions between separate jet streaks complicate the use of simple jet stream models that infer upper-level divergence and preferred sites for cyclogenesis. Sloping isentropic analyses are best suited for identifying jet streaks which extend through a significant depth of the troposphere, as Uccellini et al. (1984) demonstrated with their analyses of the February 1979 "Presidents' Day Storm."

### 5.3.2   Temporal Evolution

Large temporal changes in the strength and location of upper-level jets characterized many of the snowstorms (see Table 9). Large increases of wind speed (greater than 15 m s$^{-1}$ over a period of 24 hours or less) were observed both during and also prior to cyclogenesis along the East Coast. The speed increases were most pronounced near the crests of the upper-level ridges immediately downwind of the troughs nearing the East Coast. The 300-mb winds increased from 60 to 85 m s$^{-1}$ between 0000 UTC 29 January 1966 and 1200 UTC 29 January 1966 over the Ohio Valley and Middle Atlantic states (see section 7.9). Large increases in wind speed near the ridge crest *prior to* cyclogenesis were also shown by other cases, including January 1964 and February 1979 (see the upper-level wind analyses for these cases in sections 7.8 and 7.18). Increases in wind speed were also observed upstream of and near the upper-level trough axes both *prior to* and *during* cyclogenesis. The February 1967 and February 1983 cases exhibited this development (see the upper-level wind analyses in sections 7.11 and 7.20). The increases in wind speed prior to and during cyclogenesis are consistent with the increased geopotential height gradients and amplifying character of the upper-level troughs observed in the 20 cases. The increases of wind speed during cyclogenesis may reflect the upper-level impact of latent heating and other physical processes associated with cyclogenesis (Chang et al. 1982).

### 5.3.3   An Orientation of Jet Streaks Conducive to Heavy Snowfall

An orientation of upper-level jet streaks and their associated vertical ageostrophic circulations that is conducive to heavy snowfall in the Northeast is shown in Fig. 21. One jet is located within the upper-level trough approaching the East Coast and another

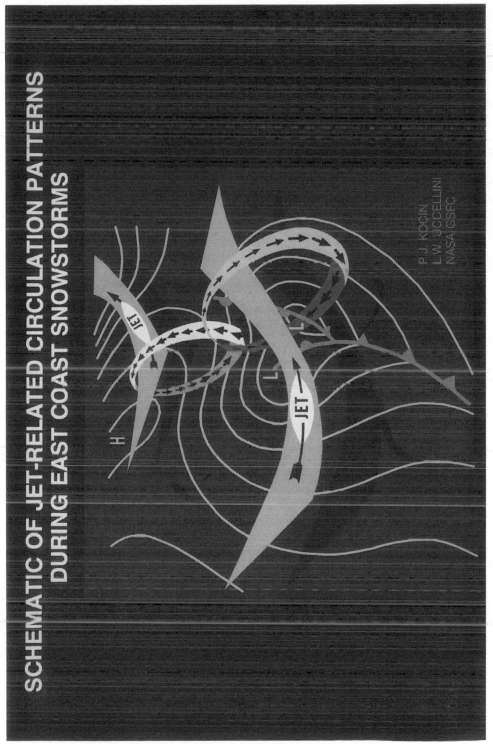

FIG. 21. Schematic of dual jet-related circulation patterns during East Coast snowstorms. Circulations are represented by pinwheels, jet streaks are imbedded within confluent and diffluent regions, and solid lines are sea-level isobars.

is situated upstream of the trough crossing southeastern Canada or the northeastern United States. An upper-level jet streak at the base of the trough, with its diffluent exit region over the East Coast, was evident in every case (Table 9) during the 12-hour period when cyclogenesis was occurring along the coast.[19] A jet streak passing across the northeastern United States or southeastern Canada was noted during the same period in at least 16 of the 20 cases (Table 9). The cold surface anticyclone or ridge line was often located beneath the region of confluent geopotential heights at 500 mb that marked the entrance region of this jet streak.

Uccellini and Kocin (1987) used eight of the cases described here to identify the configuration of the upper-level jet streaks depicted in Fig. 21. This pattern is associated with transverse vertical circulations that link the jet streaks and their associated upper-level troughs to the orientations of the surface cyclones and anticyclones, and to the temperature advections, moisture transports, and vertical motions needed to produce heavy snowfall. The analyses of these cases illustrate that:

(i) As an upper-level trough over the Ohio and Tennessee valleys approaches the East Coast and a jet streak at the base of the trough enters the diffluent region downwind of the trough axis, a surface low-pressure system develops. The cyclogenesis occurs downstream of the trough axis, within the exit region of the jet streak, where upper-level divergence (cyclonic vorticity advection) related to both the trough and the jet favors surface cyclonic development.

(ii) A separate upper-level trough over southeastern Canada and a jet streak embedded in the confluence zone over New England are associated with a cold surface high-pressure system, which is usually found beneath the confluent entrance region of the jet streak.

(iii) Indirect and direct transverse circulations that extend through the depth of the troposphere are located in the exit and entrance regions of the southern and northern jet streaks, respectively.

(iv) The rising branches of the transverse vertical circulations associated with the two jet streaks appear to merge, contributing to a large region of ascent which produces clouds and precipitation between the diffluent exit region near the East Coast and the confluent entrance region over the northeastern United States.

(v) The advection of Canadian air southward in the lower branch of the direct circulation over the northeastern United States maintains the cold lower-tropospheric temperatures needed for snowfall along the East Coast. In many cases, the lower branch of this circulation pattern is enhanced by the ageostrophic flow associated with cold air damming east of the Appalachian Mountains.

(vi) The northward advection of warm, moist air in the lower branch of the indirect circulation over the southeastern United States rises over the colder air north of the surface low. An easterly or southeasterly low-level jet streak (see section 5.4) typically develops within the lower branch of the indirect circulation beneath the diffluent exit region of the upper-level jet.

(vii) The combination of the differential vertical motions in the middle troposphere

---

[19] Corresponding to the third and fourth panels of the analyses shown in Chapters 7 and 8.

and the interactions of the lower branches of the direct and indirect circulations appears to be highly frontogenetic [as computed by Sanders and Bosart (1985a) for the February 1983 storm], increasing the thermal gradients in the middle and lower troposphere during the initial cyclogenetic period. This finding is consistent with the increase in available potential energy observed by Palmén (1951, p. 618) during the early stages of cyclogenesis.

## 5.4   LOWER-TROPOSPHERIC TROUGHS, TEMPERATURE, AND WIND FIELDS

Many weather forecasters recognize that the demarcation between rain and snow is often found with 850-mb temperatures near 0°C [other empirical approaches also stress critical values of 1000–850-mb or 1000–500-mb thicknesses; for example, see Wagner (1957)]. The 850-mb surface serves not only as an indicator of precipitation type, but is also representative of lower-tropospheric thermal and moisture advections. Thus, the 850-mb surface is a valuable level for examining the factors that influence the precipitation form and intensity during snowstorms (Bailey 1960; Browne and Younkin 1970; Spiegler and Fisher 1971).

A comparison of the 850-mb low tracks (Fig. 22) with the snowfall patterns in the 20 storms (Fig. 6) reveals that the heaviest snowfall occurred 50 to 300 km to the left of the center's path. This generally agrees with the results of Browne and Younkin (1970) and Spiegler and Fisher (1971). The paths of the 850 mb low centers were located near or to the left of their sea-level counterparts (compare Figs. 22 and 7), and generally followed east to northeasterly tracks along or off the Middle Atlantic and southern New England coasts.

Spiegler and Fisher stated that the 850-mb low center is a better guide for forecasting snow amounts than the sea-level cyclone center, since its path appears to be more continuous. A review of the 850-mb low tracks in the 20 cases, however, shows that the movement of these centers may have been as discontinuous as the paths of the surface low-pressure centers that redeveloped along the coast (top two panels of Fig. 22; also see Table 10). When the redeveloping surface low-pressure systems intensified along the coast (those cases shown in the top two panels of Figs. 7 and 22), the associated changes in the lower troposphere were displayed at the 850-mb level as either a distinct secondary low center or as a separate maximum of geopotential height falls. It is difficult to properly monitor the secondary development at 850 mb due to the 12-hour interval between the operational radiosonde measurements, which cannot adequately resolve the rapidly evolving 850-mb troughs during cyclogenesis (e.g., see Uccellini et al. 1987).

The 850-mb charts for the 20-case sample also revealed that during a 12- to 36-hour period preceding cyclogenesis, northwesterly flow and cold air advection usually covered the Northeast, as depicted schematically in Figs. 23a and 23b. The cold air advection occurred to the rear of an intense 850-mb low centered over southeastern Canada or the adjacent offshore waters. Northwesterly flow with strong low-level cold air advection was observed in at least 15 of the 20 cases (see Table 10), driving the

**PATHS OF GEOPOTENTIAL HEIGHT MINIMA AT 850 MB**

FIG. 22.   Paths of geopotential height minima at 850 mb. The 20 cases are grouped as in Fig. 7.

0°C isotherm south of the urban centers of the northeastern United States toward the Carolinas. The northwesterly flow was associated with the surface anticyclones that influenced the northeastern United States (Fig. 12), and was generally located beneath the regions of upper-level confluence (see section 5.2 and Table 8).

Another characteristic of the 20-case sample is the evolution of a geopotential height minimum at 850 mb along a preexisting isotherm gradient extending from the central United States to the East Coast (Fig. 23). The 850-mb height minimum, or low center, intensified during cyclogenesis. Meanwhile, the 850-mb temperature gradients increased and evolved into an "S"-shaped pattern, as the advections of cold and warm air attending the 850-mb circulation increased with time (Figs. 23b and 23c). Prior to rapid cyclogenesis, the 0°C isotherm progressed northward toward the Middle Atlantic states and became nearly collocated with the 850-mb low center. During cyclogenesis, the 0°C isotherm did not progress rapidly northward despite increasing south or southeasterly winds. This implies that cooling resulting from strong ascent may have offset the effects of the thermal advection (Browne and Younkin 1970). The sea-level cyclone was often located along the warm edge of the 850-mb temperature gradient, either at or upwind of the apex of the isotherm ridge (see sea-

TABLE 10.  Characteristics of 850-mb geopotential height, temperature, and wind fields, including the occurrence of dual 850-mb lows or geopotential height fall centers, the occurrence of northwesterly flow/cold air advections over the northeastern United States during a 36-hour period preceding cyclogenesis along the East Coast, evidence of "cold air injections" (see text) to the rear of the 850-mb low center, and maximum observed wind velocity (m s$^{-1}$; direction in parentheses) in the low-level jet along the East Coast.

| Date | Dual 850-mb or geopotential height fall centers | Northwesterly flow/cold air advection prior to cyclogenesis | "Cold air injections" to the rear of the 850-mb low | Maximum observed wind speed (m s$^{-1}$) and direction in the low-level jet |
|---|---|---|---|---|
| 18–20 Mar 1956 | X | X | X | 25 (E) |
| 14–17 Feb 1958 | — | X | X | 25 (E) |
| 18–21 Mar 1958 | — | — | X | 35 (ENE) |
| 2–5 Mar 1960 | X | X | X | 30 (ESE) |
| 10–13 Dec 1960 | — | X | X | 25–30 (NE) |
| 18–20 Jan 1961 | — | X | X | 30 (SSW) |
| 2–5 Feb 1961 | X | X | X | 30 (SE) |
| 11–14 Jan 1964 | X | X | X | 30 (E) |
| 29–31 Jan 1966 | — | X | X | 30 (S) |
| 23–25 Dec 1966 | — | X | X | 25–30 (E) |
| 5–7 Feb 1967 | — | X | X | 30 (SE) |
| 8–10 Feb 1969 | X | weak | X | 30–35 (SE) |
| 22–28 Feb 1969 | — | — | — | 30 (SE) |
| 25–28 Dec 1969 | — | weak | X | 30–35 (E) |
| 18–20 Feb 1972 | X | — | X | 30 (SE) |
| 19–21 Jan 1978 | X | X | X | 25–30 (E) |
| 5–7 Feb 1978 | X | X | X | 35 (E) |
| 18–20 Feb 1979 | X | X | — | 25 (SE) |
| 5–7 Apr 1982 | X | X | X | 25–30 (NE) |
| 10–12 Feb 1983 | — | X | — | 20 (NE) |

level low-pressure center positions also depicted in Fig. 23). As the cyclone occluded, the 850-mb low frequently became nearly collocated with the sea-level cyclone in an area with 850-mb temperatures below 0°C (not shown).

## 5.4.1  Low-Level Jet Streaks

A pronounced 850-mb isotach maximum or low-level jet streak (LLJ) is an important feature that develops either prior to or during cyclogenesis. The 850-mb analyses in Chapter 7 illustrate that several lower-tropospheric wind speed maxima occurred during different stages of cyclogenesis in most of the cases. One type of LLJ has been termed a "cold air injection." It is directed from the north or northwest behind the 850-mb low center, and is typically found over the central or south-central United States (Bailey 1960). This jet developed prior to cyclogenesis along the East Coast and was usually oriented normal to the isotherms, maximizing cold air advection over the south-central United States (Figs. 23a and 23b). While this feature has been identified as a clue to subsequent cyclogenesis (Bailey 1960), little physical explanation has been

**REPRESENTATION OF 850 MB SURFACE FOR A HEAVY SNOW EVENT
ALONG THE NORTHEAST COAST**

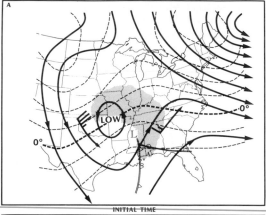

INITIAL TIME

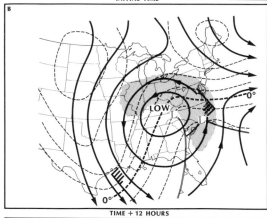

TIME + 12 HOURS

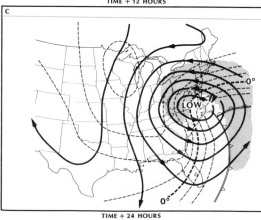

TIME + 24 HOURS

Fig. 23.   Representation of the 850-mb surface at 12-hour intervals for a heavy snow event along the Northeast coast, including streamlines (solid arrows), isotherms (dashed), wind barbs {indicating wind speeds and directions near "cold air injections" and low-level jet streaks [each flag represents 25 m s$^{-1}$; each barb represents 5 m s$^{-1}$; each half-barb represents 2.5 m s$^{-1}$ (see text)]}, and sea-level low-pressure center and surface frontal positions (open lines).

offered for its significance. In this 20-storm sample, 17 cases clearly displayed this cold air injection (Table 10), defined as 850-mb winds in excess of 15 m s$^{-1}$ crossed an isotherm ribbon with temperature gradients of 12°C (5° latitude)$^{-1}$ or greater to the rear of an 850-mb low center.

With the onset of rapid cyclogenesis along the East Coast, a separate LLJ developed from a southerly to easterly direction near the coast (Table 10). This LLJ was observed in all 20 cases, with its maximum wind speeds typically ranging between 25 and 35 m s$^{-1}$ (Table 10). The development of wind maxima near the 850-mb level during East Coast storms may account for the secondary maximum of LLJs frequently found along the North Carolina coast by Bonner (1965) in his climatological survey of low-level jet streaks in the United States (the primary maximum is located over the southern Plains states).

While the LLJ usually developed concurrent with the deepening of the 850-mb vortex (for example, see Fig. 24a), it sometimes formed near the coast *in advance* of the development of a separate 850-mb low center (Fig. 24b). The rapid development of these low-level jet streaks occurred in cases where dual 850-mb geopotential height minima were observed in association with secondary surface cyclogenesis (Table 10), and were most pronounced in the February 1979 and January 1964 cases (see the 850-mb analysis in sections 7.18 and 7.8).

One such LLJ, with wind speeds approaching 25 m s$^{-1}$, formed over Georgia and South Carolina between 0000 and 1200 UTC 18 February 1979, prior to the "Presidents' Day Storm" (Fig. 24b). The development occurred in an area of significant geopotential height falls displaced a considerable distance from the primary 850-mb trough, which was located over the center of the country. The LLJ developed within a 4-hour period, as shown in a model-based study by Uccellini et al. (1987), in response to the upper-level divergence associated with a subtropical jet streak, the sensible heat exchange in the ocean-influenced boundary layer, and coastal frontogenesis. The study cited above illustrates that these physical and dynamical processes interact to form the southerly to southeasterly wind maxima that enhance the temperature advection and moisture transport toward the regions of heaviest snowfall (Uccellini et al. 1984). The rapid development of the LLJ can also amplify the mass divergence along the coastline, focusing the sea-level pressure falls that mark secondary cyclogenesis along the coast (Uccellini et al. 1987). Thus, it appears that the development of the LLJ makes an important contribution to thermal and moisture advections and mass divergence that focus the secondary cyclogenesis along the Southeast coast.

## 5.5 SATELLITE IMAGERY AND PERSPECTIVE OF THREE-DIMENSIONAL AIRFLOW

Satellite-derived images of cloud patterns, especially the half-hourly high-resolution visible and infrared images provided by the Geostationary Operational Environmental Satellite (GOES) series, have helped to visualize the evolution of cloud elements and the three-dimensional airflow associated with heavy snowstorms along the Northeast coast. Meteorologists who analyze satellite data have long recognized that the "comma-

## TWO ILLUSTRATIONS OF LOW-LEVEL JET FORMATION AT 850 MB

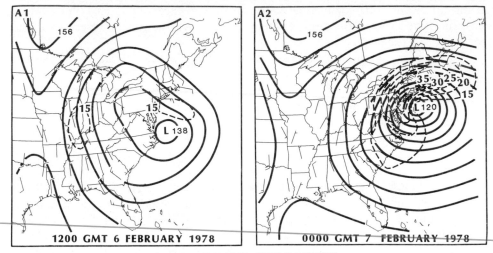

**DURING RAPID CYCLOGENESIS**

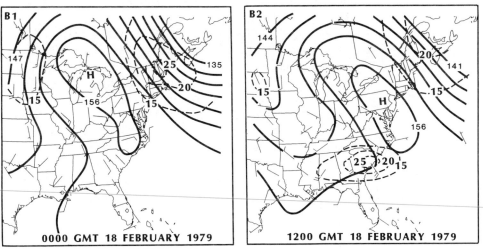

**PRECEDING RAPID CYCLOGENESIS**

FIG. 24.   Illustrations of low-level jet formation at 850 mb at (A1, A2) 1200 UTC 6 February to 0000 UTC 7 February 1978 and (B1, B2) 0000 to 1200 UTC 18 February 1979, including geopotential height (solid, 138 = 1380 m), isotachs (dashed, m s$^{-1}$), and wind barbs (each barb represents 10 m s$^{-1}$; each half-barb represents 5 m s$^{-1}$).

cloud" pattern is a signature of these storms (for example, see Boucher and Newcomb 1962; Leese 1962; and Widger 1964), as shown by the visible and infrared GOES images for the five most recent cases in Fig. 25.[20] While there were distinct case-to-case differences in the orientations of these cloud features, the comma-shaped appearance characterized each case.

[20] Continuous visible and infrared imagery are available only since 1974. Therefore, complete satellite documentation for the first 15 cases does not exist.

A conceptual model of midlatitude cyclones that relates the comma-cloud signatures apparent in satellite imagery to the relative airflow through the storm systems was described by Carlson (1980). This model was based on trajectory analyses on constant potential temperature, or isentropic, surfaces in dry regions and on wet-bulb potential temperature surfaces in saturated regions (Eliassen and Kleinschmidt 1957; Green et al. 1966; Browning and Harrold 1969; Browning 1971), and identifies three airstreams associated with cyclones: the "warm conveyor belt," the "cold conveyor belt," and the "dry airstream." A schematic diagram that illustrates the primary components of the model in relation to a hypothetical heavy snow situation in the Northeast is shown in Fig. 26.

The warm conveyor belt (Harrold 1973; Browning 1986) is a stream of air that originates in the warm sector of a cyclone and ascends as it moves northward, achieving saturation near the warm front. It then turns anticyclonically and joins the upper-level westerly flow north of the sea-level cyclone (see Fig. 26). While isobaric charts are incapable of capturing the substantial vertical displacements that characterize this and the other airstreams, the warm conveyor belt may be represented on the surface charts by the southerly flow ahead of the surface cold front, and on the upper-level charts (i.e., the 500-, 300-, and 200-mb analyses) by the southwesterly flow downwind of the trough axis. In general, however, the warm conveyor belt may be difficult to diagnose in cyclones that yield heavy snow along the Northeast coast since it is located primarily over the data-sparse Atlantic Ocean.

The warm conveyor belt typically forms the comma "tail" and may be bounded by a distinct anticyclonically curved shield of cold upper-level cirrus clouds as the airstream ascends and veers with height [see Carlson (1980) and the model-diagnostic study by Whitaker et al. (1988)]. The 20 January 1978 and 6 February 1978 visible and infrared images in Fig. 25 illustrate the cloud elements associated with this airstream (labeled WCB in the infrared imagery). An examination of the complete set of satellite observations from the five storms displayed in Fig. 25 shows that only the January 1978 case had the warm conveyor belt over the region of heavy snowfall during some portion of the cyclone's evolution.

The cold conveyor belt (Carlson 1980) may be an especially significant component of snowstorms that affect the Northeast coast. It originates within the anticyclonic flow around the surface high-pressure cell poised to the north of the developing storm system. The air within this stream moves westward within the ocean-influenced boundary layer, toward and to the north of the surface low. This airstream then ascends the steeply sloped isentropic surfaces associated with the upper-level trough/jet system approaching the East Coast (Fig. 26), with a portion of the airstream circulating around the surface low. If air parcels ascend to a high enough level, the cold conveyor belt may also turn sharply to the north or northeast as it rises (Carlson 1980). The model-based trajectories shown by Whitaker et al. (1988; cf. their Fig. 18, parcel #5) provide evidence for this characteristic. The cold conveyor belt is viewed on the operational charts as the surface flow around the southern periphery of the shallow, cold anticyclone north of the warm front, and the easterly flow north of the 850-mb low center. Saturation in the cold conveyor belt results from precipitation falling into it from above as it passes beneath the warm conveyor belt, moisture fluxes from the ocean surface,

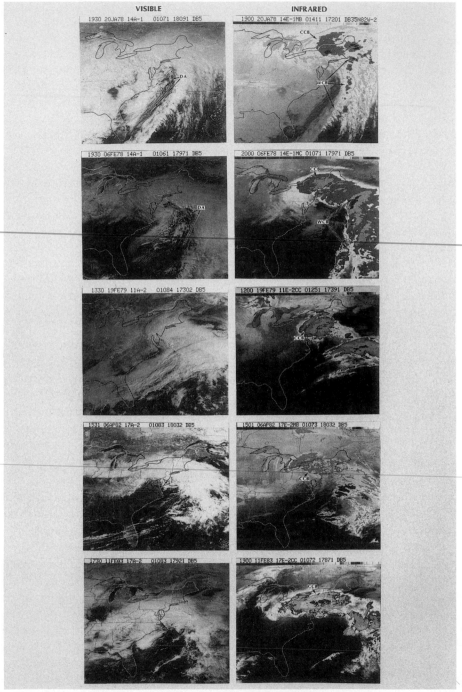

Fig. 25. Geostationary Operational Environmental Satellite (GOES) visible (left) and infrared (right) images of "comma"-shaped cloud patterns during five major Northeast snowstorms. First two alphanumeric groups at the top left of each image indicate the time (UTC) and date (20JA78 = 20 January 1978). Abbreviations represent the following: CCB (cold conveyor belt), WCB (warm conveyor belt), and DA (dry airstream); from Carlson (1980). Gray scales in the infrared imagery correspond to different ranges of temperature, with the enhanced areas indicating colder (higher) cloud tops.

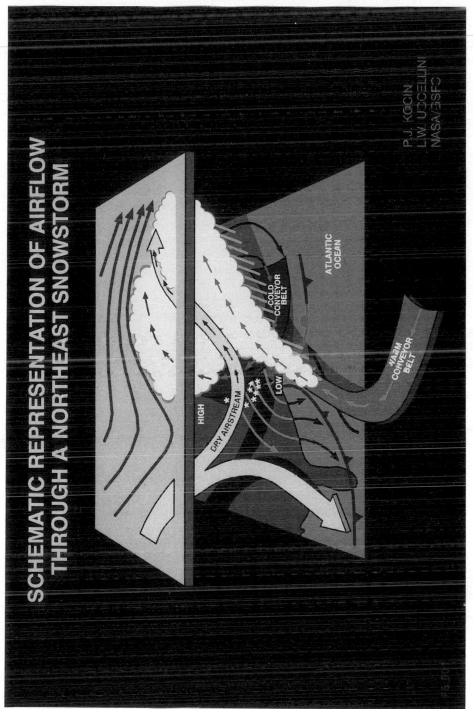

Fig. 26. Schematic representation of airflow through a Northeast snowstorm, including the cold conveyor belt, the warm conveyor belt, the dry airstream [see Carlson (1980)], surface fronts, sea-level cyclone and anticyclone centers, and low- and upper-level streamlines.

and the rapid ascent of the airstream as it passes to the north of the developing cyclone center.

The cold conveyor belt produces an upper-level cirriform cloud deck that forms the distinctive comma "head" north of the sea-level cyclone, which generally expands westward or northwestward with time. The visible and infrared GOES images for all five cases in Fig. 25 display this airstream, labeled CCB in the infrared images. The cold conveyor belt is quite important, since heavy snowfall occurred within the expanding comma-head cloud mass in each of the five cases with complete satellite documentation. Major snow amounts were observed near the transition between the high cloud tops that mark the comma head and the more shallow clouds immediately to its south and southwest. Similar relationships between the location of heavy snowfall and the distribution and height of the clouds, as inferred from satellite imagery, have been reported by Carlson (1980), Uccellini et al. (1984), Bosart and Sanders (1986), Beckman (1987), and others. No precipitation was reaching the Earth's surface under a significant portion of the cirrus cloud deck that made up the comma head, especially poleward of the highest clouds depicted by the enhancement in the infrared imagery. These nonprecipitating cloud elements suggest that a relatively thin layer of high clouds was produced by the ascent of the airstream and was then advected northeastward with the upper-level flow.

The dry airstream contains dry air that originates in the lower stratosphere west of the trough axis and descends along an upper-level jet streak toward the developing surface cyclone. Many investigators (Eliassen and Kleinschmidt 1957; Danielsen 1966; Uccellini et al. 1985; Hoskins et al. 1985; Whitaker et al. 1988) have linked the descent of stratospheric air to mesoscale circulation patterns that produce a potential-vorticity maximum upstream of the incipient cyclone, a situation that is conducive to cyclogenesis. The descending airstream splits into two branches. One branch descends anticyclonically toward the lower troposphere behind the surface cold front. The other descends initially upstream of the trough axis, then crosses the trough axis, and rises roughly parallel to the western edge of the warm conveyor belt (see Fig. 26). This intrusion of dry air is relatively cloud-free and helps to shape the comma cloud, as it appears to be pinching off the comma head from the tail (e.g., see the visible images for 20 January 1978 and 6 February 1978 in Fig. 25, where the dry airstream is labeled DA). The interaction of the dry airstream and the moisture-laden warm and cold conveyor belts is largely responsible for the asymmetric cloud distribution that marks these storms. As shown in Fig. 26, the surface low is generally located along the boundary between the dry airstream approaching from the west and the moist warm and cold conveyor belts approaching from the south and east.

In the later stages of cyclone development, when the systems have moved out over the Atlantic Ocean, satellite images often reveal the formation of an "eye"-like cloud-free region near the storm centers, similar to that observed in tropical cyclones (also described by Sanders and Gyakum 1980; Bosart 1981; and Uccellini et al. 1985). Visible satellite imagery from the February 1969 "Lindsay Storm" (see section 7.12) and from four of the five most recent cases are displayed in Fig. 27 to illustrate the appearance of this feature during or toward the end of each cyclone's rapid development

FIG. 27. Applications Technology Satellite (ATS-III) and GOES visible imagery of "eye"-like cloud features associated with Northeast snowstorms. First two alphanumeric groups at the top left of each image represent the time (UTC) and the date (09FE69 = 9 February 1969).

phase. An examination of both visible and infrared imagery for the four more recent cases shown in Fig. 27 indicates that the height distribution of the clouds that surround the clear centers are asymmetric. Deeper clouds were observed to the north, with relatively shallow clouds south of and immediately surrounding the storm center. Although convective elements may have been present, the asymmetric cloud height distribution did not resemble the deep convective cloud patterns that typically surround the centers of mature, intense tropical cyclones.

CHAPTER **6**

# Summary of the Physical and Dynamical Processes That Influence Northeast Snowstorms

Twenty severe snowstorms that affected the northeastern coast of the United States between 1955 and 1985, have been selected for study based on their extensive distribution of heavy snowfall (>25 cm) across one of the most heavily populated and industrialized regions of the world, including the metropolitan areas of New York, Washington, Boston, Philadelphia, and Baltimore. Analyses of the surface and upper-level fields of pressure, geopotential height, wind, and temperature, plus other selected elements described in the preceding two chapters, provide a framework for understanding the dynamical and thermodynamical processes that contribute to heavy snowfall in the northeastern United States. These analyses point out the important role of cyclogenesis, which was observed to some degree in each of the 20 snowstorms. Surface cyclogenesis involved either a primary low-pressure center or a secondary development along the Southeast or Middle Atlantic coast that tracked northeastward approximately 100 to 300 km offshore.

Cyclogenesis encompasses a wide range of processes that organize the vertical motions necessary for heavy precipitation. Upper-level trough/ridge systems characterized by changing amplitude, wavelength, and orientation, and multiple jet streaks at various levels of the troposphere each made a significant contribution to the cyclogenesis that accompanied the 20 cases. Cyclogenesis requires these favorable upper-tropospheric factors, but the snowstorms also need conditions that enhance low-level baroclinicity near or immediately off the coastline and that maintain cold air over the coastal plain so that the major metropolitan areas experience snow rather than rain. An environment conducive to heavy snowfall is usually not provided solely by the presence of a cyclone, but also requires its interaction with an anticyclone of Canadian origin.

The evolution of cyclonic and anticyclonic weather systems that interact to produce major coastal snowstorms is influenced by the complex topography of the eastern United States, which features a narrow coastal plain bounded by the Appalachians to the west and the Atlantic Ocean to the east. Topographically influenced physical processes in the lower troposphere can enhance low-level baroclinicity near the coastline, forming a "coastal front," and can maintain cold air just west of the coastline through cold air damming. The Atlantic Ocean and its Gulf Stream readily supply warmth and moisture to augment the coastal baroclinicity and to support the development of heavy precipitation over a large area. The processes that are conducive to heavy snowfall in the northeastern United States will be reviewed in the following sections.

## 6.1    REVIEW OF DYNAMICAL AND PHYSICAL PROCESSES

Table 11 and Fig. 28 provide a written and visual summary of the factors that contribute to heavy snowfall along the Northeast coast of the United States. The following brief descriptions of the relevant dynamical and physical processes show how each influences the development of these storms. Although the processes were treated separately to identify their individual contributions, we emphasize that the nonlinear interaction of these processes is crucial for the development of heavy snowfall. This and other remaining issues that are presently not well understood are addressed at the end of the chapter.

### 6.1.1    Upper-Level Troughs

The fundamental contributions of upper-level troughs and the jet stream to the upper-level divergence necessary for surface cyclogenesis were emphasized by Bjerknes and Holmboe (1944), Sutcliffe (1939, 1947), Charney (1947), Bjerknes (1951), Petterssen (1956), Palmén and Newton (1969), and many others. Each heavy snow/cyclogenesis event was associated with a well-defined, preexisting upper-level trough and cyclonic vorticity advection pattern. Although the presence of an upper-level trough was common to every storm, the study indicates that these systems evolved in many different ways.

    Most of the systems evolved from an open-wave structure into a vortical circulation at 500 mb (12 of the 20 cases). Some others remained as open-wave troughs (4 cases), and the rest exhibited a vortical structure both prior to and during cyclogenesis (4 cases). Although the evolution of the trough/ridge configuration varied from one storm to another, 18 of the 20 cases displayed an increase of amplitude, and all 20 cases showed a decreasing half-wavelength between the trough and downstream ridge axes during cyclogenesis. Both of these characteristics are indicative of nonlinear processes that increase upper-level divergence during surface cyclogenesis [e.g., see Sutcliffe and Forsdyke (1950) and Palmén and Newton (1969, pp. 324–326)]. A marked diffluence of the upper-level height contours downstream of the trough axis was displayed in all 20 cases, which is another signature of increasing upper-level divergence (Scherhag 1937; Bjerknes 1954; Palmén and Newton 1969, pp. 334–335). This pattern was especially pronounced during the period of most rapid cyclogenesis and the development of heavy snowfall. The diffluence of the height contours increased dramatically as the trough axes assumed a northwest to southeast orientation, or negative tilt.

    All of the upper troughs contained one or more cyclonic vorticity maxima, pro-

TABLE 11.    Meteorological phenomena involved in the development of East Coast snowstorms. Scales are defined as synoptic (24 to 48 hours; ≥2000 km) and large mesoscale or meso-α (3 to 24 hours; 100 to 2000 km).

| Meteorological phenomena | Scale | Important processes for the development of Northeast snowstorms |
|---|---|---|
| 6.1.1 Upper-level trough/ridge pattern | synoptic | Provides source of vorticity and kinetic energy, upper-level divergence, and associated ascent. Amplification of this feature often marked by decreasing wavelength, increasing amplitude, and the development of diffluence downwind of a "negatively tilted" trough axis. |
| 6.1.2 Upper-level jet streaks | meso-α | Increase upper-level divergence, kinetic energy, and vorticity. Combination of jet streaks could be an important factor in juxtaposing different air masses and enhancing ascent patterns required for heavy snows. |
| 6.1.3 Cold anticyclone extending across the eastern Great Lakes, southern Canada, or New England north of the incipient low | synoptic | Provides cold air required to maintain snow along the coast, a process augmented by cold air damming east of the Appalachian Mountains. |
| 6.1.4 Processes in the lower troposphere, flow interaction with the Appalachian Mountains, cold air damming, and coastal frontogenesis | meso-α (and smaller) | Focus low-level baroclinic zone, ascent, and temperature advections in a narrow region along the coast. Coastal front enhances precipitation rates. |
| Low-level jet | meso-α | Increases the transport of moisture into region of heavy snow, warm air advection, and associated ascent pattern. Plays a possible role in focusing mass-flux divergence during cyclogenesis along the coast. |
| Sensible and latent heat fluxes in the oceanic PBL | meso-α (and smaller) | Warms the lower troposphere by reducing the static stability of planetary boundary layer over the ocean; provides a continuing source of moisture for heavy precipitation over a large area. |
| 6.1.5 Latent heat release | meso-α | Provides extra heat for intensifying cyclogenesis through the self-development process. |

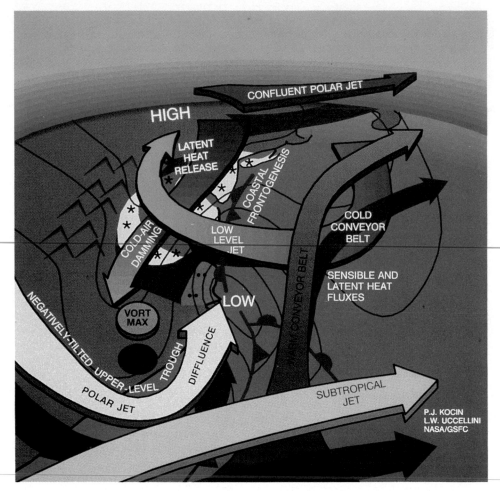

FIG. 28.   Schematic of factors that contribute to heavy snowfall along the northeastern coast of the United States.

ducing the vorticity advection that is commonly used to infer upper-level divergence. Although the vorticity advections were not directly evaluated, the increasing amplitudes and decreasing wavelengths of the trough/ridge systems suggest that cyclonic vorticity advection was increasing over the East Coast during cyclogenesis in many, if not all, cases. The vorticity maxima tracked eastward or northeastward across the Middle Atlantic states in most of the cases. Heavy snowfall is much less likely when the vorticity maxima propagate to the west or north of the region since the lower-tropospheric warm air advection associated with these situations favors rain, freezing rain, or ice pellets rather than snow. The merging or phasing of multiple vorticity maxima into one cohesive maximum was noted in nearly half the cases. The merger may have strengthened the cyclonic vorticity advection and upper-level divergence, thereby increasing cyclone development. The dilemma in determining whether separate trough and vorticity features will merge remains a critical concern to forecasters who must predict surface cyclone development and heavy snowfall using numerical guidance, which displays varying degrees of skill in predicting these interactions.

### 6.1.2    Upper-Level Jet Streaks

In contrast to the relatively straightforward characterizations of the upper-level troughs in the 20-storm sample, the structure and evolution of jet streaks during individual cyclonic episodes proved to be very complicated. The presence of jet-like, finite length wind maxima is known to enhance upper-level vorticity advections and divergence, which contribute to surface cyclogenesis [e.g., see Bjerknes (1951), Palmén and Newton (1969), and Uccellini and Kocin (1987)]. A jet streak was evident at the base of the upper-level trough nearing the East Coast in every case, with the surface cyclogenesis occurring in the diffluent exit region of the jet. There are often several other jet streaks during each heavy snow episode, however, some at different levels of the atmosphere. The evolution of these systems is often difficult to diagnose using the 12-hourly operational radiosonde data. Large temporal changes in wind speed and in the vertical and horizontal location of the jet core occurred in many of the cases. These problems in assessing the evolution of multiple jet streaks restrict the use of simple conceptual models that relate cyclogenesis to the patterns of upper-level divergence in the left-front exit and right-rear entrance regions of a single straight jet streak (see Riehl et al. 1952).

The jet-related divergence patterns and their associated circulations described in the conceptual models have been identified in a significant number of Northeast snowstorms by Uccellini and Kocin (1987). They indicated that the combination of a jet streak at the base of an upper-level trough over the southeastern United States and a separate polar jet streak over the northeastern United States or southeastern Canada enhance the ascent north of the surface low. This scenario, which is illustrated in Fig. 21, places a transverse indirect circulation within the exit region of the jet streak associated with the upper-level trough approaching the East Coast. A direct transverse circulation is located within the entrance region of the jet streak associated with the upper-level confluence zone over southeastern Canada. The advection of cold Canadian air southward along the coast is associated with the lower branch of the direct circulation. The northward advection of warm, moist air into the region experiencing heavy precipitation occurs in the lower branch of the indirect circulation. Thus, the jet-streak-induced circulations not only enhance the ascent necessary for heavy precipitation, but also contribute to the horizontal temperature and moisture advection patterns which juxtapose the cold air and the warm, moist air in the manner required for heavy snowfall.

### 6.1.3    Cold Surface Anticyclones

Anticyclones that pass across southeastern Canada or surface high-pressure ridges that extend eastward from anticyclones over the American or Canadian plains provide the cold low-level air that enables precipitation to reach the ground as snow. These anticyclones have Canadian or polar origins, maintain central pressures in excess of 1030 mb, and remain in a position to reinforce the advection of cold air over the coastal plain. The surface anticyclone remains generally to the north of the surface low, resulting in a low-level flow of cold air which is minimally influenced by the warming effects of the nearby ocean. The displacement of an initially cold surface anticyclone too far to the east produces a flow of ocean-warmed air toward the northeastern United States, often resulting in rainfall along the immediate coast. While the presence of an anticyclone to the north should not be viewed as a prerequisite for heavy snowfall (e.g., the 9–10 February 1969 storm followed an anticyclone that moved eastward off the coast prior to cyclogenesis), it is a common feature of nearly all of the cases presented here.

As a general rule, the position of the anticyclone relative to the evolving storm system is a key factor in driving cold air toward the coastal region, with minimal modification due to its primarily over-land trajectory. The maintenance of a cold anticyclone to the north of the developing low is linked to a consistent pattern of geopotential heights in the northeastern United States and southeastern Canada. In all 20 cases, the anticyclone was located upwind of an upper-level trough crossing eastern Canada, almost always beneath a region of confluent geopotential heights (19 cases), and usually within the entrance region of an upper-level jet streak (16 cases). The confluent flow pattern within the entrance region of the jet streak over southeastern Canada maintains the surface anticyclone, enhances the advection of cold air southward along the coastal plain, and, therefore, plays a critical role in the development of most of these snowstorms (Uccellini and Kocin 1987).

### 6.1.4   Processes in the Lower Troposphere

While upper-level troughs, jet streaks, and vorticity advection contribute to the formation and development of the storms, physical processes in the lower troposphere also play important roles in creating the conditions that generate heavy snowfall in the northeastern coastal region. Charney (1947), Sutcliffe (1947), Petterssen (1955), and many others have shown that thermal advection patterns below 700 mb are important components of cyclogenesis, affecting surface deepening rates, vertical motions, and the energy conversions associated with developing cyclones. The contribution of thermal advections to the vertical motions associated with cyclogenesis can be assessed from the "development equations" of Sutcliffe (1947) and Petterssen (1955), and from the more recent quasi-geostrophic framework (i.e., Holton 1979, Chapter 7), which has been used extensively in diagnostic studies of cyclogenesis.

The lower-tropospheric thermal advection patterns associated with snowstorms along the northeastern coast of the United States are strengthened within the baroclinic zone along the East Coast. Every storm in the 20-case sample developed in association with a preexisting isotherm gradient at the 850-mb level, which intensified and evolved into an S-shaped configuration. A circulation center developed, near the coast, above and slightly to the west of the surface low-pressure center, close to the inflection point of the S-shaped isotherm configuration.

The lower-tropospheric thermal advections surrounding the 850-mb low center increased during cyclogenesis. Part of this increase was attributable to the effects of topography on the boundary-layer temperature structure. The Appalachian Mountains, the coastline, and the warm waters of the Atlantic Ocean and the Gulf Stream are particularly important factors. Although the Appalachians may facilitate cyclone development through "vortex-tube" stretching as air flows over the mountains (i.e.,

Holton 1979, pp. 87–89; Smith 1979, pp. 163 169), they probably make a more significant contribution by entrapping cold air between the mountains and the coastline. This process, termed cold air damming (Richwien 1980; Forbes et al. 1987; Stauffer and Warner 1987), is marked by a distinctive inverted surface high-pressure ridge and was observed in 19 of the 20 cases. It occurs when cold air supplied by an anticyclone over New England or Canada is channeled southward along the eastern slopes of the Appalachians. Damming contributes to the southward advection of cold air, and is maintained by a combination of ascent, evaporation, and orographic, frictional, and isallobaric effects (Forbes et al. 1987; Bell and Bosart 1988).

The cold air advection associated with damming combines with the fluxes of sensible and latent heat from the ocean surface that warm and moisten the overlying atmosphere (Petterssen 1962; Sanders and Gyakum 1980; Bosart 1981; Gyakum 1983a,b) to intensify the thermal contrast near the coastline. This process, known as coastal frontogenesis (Bosart et al. 1972; Bosart 1975; Ballentine 1980; Bosart 1981; Bosart and Lin 1984; Forbes et al. 1987; Stauffer and Warner 1987), is confined to the lower troposphere, typically precedes cyclogenesis, and focuses low-level conver gence, baroclinicity, warm air advection, and surface vorticity. Coastal frontogenesis was observed in 16 of the 20 cases studied, and is an integral component of the sequence of processes that interact to produce heavy snowfall along or just west of the coast.

The intensification of the low-level thermal advections is also connected with the development of LLJs directed at a significant angle to the isotherms. The LLJ, which was observed in all 20 cases, was directed primarily from an east-southeasterly direction, with maximum wind speeds generally between 25 and 35 m s$^{-1}$ at 850 mb. The LLJ is often directed from the oceanic planetary boundary layer toward the coastal plain, where it rises over the cold air mass established between the Appalachians and the coastline. In addition to increasing the warm air advections and moisture transports during cyclogenesis, the LLJ also contributes to surface pressure falls by increasing the net divergence above the developing surface low along the coast (Uccellini et al. 1987).

### 6.1.5   Latent Heat Release

Danard (1964) recognized that the inability of the first numerical models to accurately predict cyclogenesis was partially due to their neglect of latent heat release. He found it necessary to include latent heat release in the model simulations to account properly for the distribution and magnitude of the ascent pattern associated with cyclogenesis and the deepening rate of the storms. This was verified by Krishnamurti's (1968) application of the quasi-geostrophic "ω-equation" to the study of the life cycle of a rapidly intensifying cyclone. Diagnostic work by Johnson and Downey (1976) also showed that latent heat release is a major contributor to the enhancement of mass circulations and the associated angular momentum transports required for cyclogenesis. A modeling study by Chang et al. (1982) illustrated the impact of latent heat release on the deepening rate of the surface cyclone and on the structure of the accompanying upper level trough/ridge system and its related wind fields.

Nearly all of the 20 cases exhibited a period of rapid sea-level development along the East Coast or over the Atlantic Ocean, which coincided with the occurrence of heavy precipitation to the north and west of the deepening cyclone. Assessing the importance of latent heat release relative to the contributions of the dynamical processes described earlier is difficult, but the simultaneous observation of rapid deepening rates and heavy precipitation indicates that latent heat release probably has a significant impact on the development of these storms.

### 6.1.6    Self-Development Processes

Although the dynamical and physical processes described above are necessary elements in the development of many cyclones, each process acting alone would not be sufficient to generate a major cyclone, let alone guarantee that precipitation would fall as snow along the coastal plain. While the problem of identifying necessary conditions for heavy snowfall remains unresolved, the mechanisms that interact to promote cyclogenesis are clues to the conditions that must exist to produce heavy snow.

The self-development concept defined by Sutcliffe and Forsdyke (1950) and discussed by Palmén and Newton (1969) provides a basis for describing the interactions between dynamical and physical processes that contribute to the development of cyclones and their associated heavy snowfall. The self-development scenario can be applied to Northeast snowstorms, as depicted schematically in Fig. 29. The approach of an upper-level trough/jet system toward the East Coast and the initial development of a surface cyclone focus the effects of warm air advection, sensible heat and moisture fluxes in the PBL, and latent heat release above 850 mb, thereby warming the lower and middle troposphere north and east of the surface low. This warming slows the eastward progression of the downstream upper-level ridge and builds its amplitude, while the trough moves eastward, amplifies, and attains a negative tilt. The decrease in the wavelength between the trough and ridge axes, an increase in the diffluence downstream of the trough, and the corresponding increase in the maximum wind speeds of the upper-level jet streaks all combine to strengthen the divergence in the middle to upper troposphere above the surface low. The increased upper-level divergence, as indicated by an enhancement of the cyclonic vorticity advection, produces further deepening of the cyclone.

As the surface low deepens, the lower-tropospheric winds surrounding the storm strengthen, especially north and east of the low, where isallobaric effects and vertical motions help to accelerate air parcels (Uccellini et al. 1987; Whitaker et al. 1988) and frictional effects are minimal. An LLJ develops to the north and/or east of the storm center (Fig. 29b), intensifying both the warm air advection pattern and the moisture transports into the region of heavy snowfall. The development of the LLJ enhances the warming beneath the ridge, further contributing to the self-development process.

The self-development process is also augmented by the mesoscale factors associated with cold air damming, sensible and latent heating within the ocean-influenced PBL, and coastal frontogenesis. Therefore, the interaction between the synoptic and mesoscale processes listed in Table 11 is crucial, as an upper-level trough and its associated jet streaks approach the highly baroclinic environment along the East Coast. These complex interrelationships between dynamical and physical processes occasionally manifest themselves in the form of major snowstorms affecting the major cities of the Northeast.

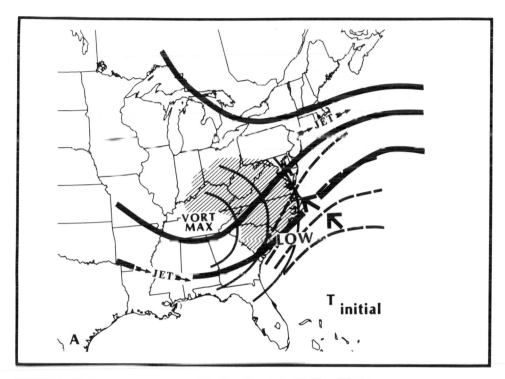

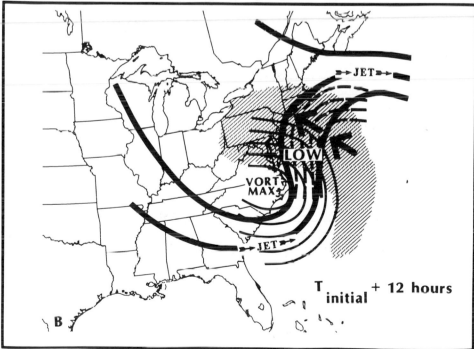

FIG. 29.   Schematic of the "self-development" concept that describes how the temperature advections, sensible heat fluxes and moisture fluxes in the PBL, and latent heat release associated with cyclogenesis enhance the amplitude of upper-level trough/ridge systems and decrease the wavelength between trough and ridge axes, which contributes to an increase in upper level divergence that further enhances cyclogenesis.

## 6.2   REMAINING ISSUES AND FUTURE RECOMMENDATIONS

This monograph demonstrates that conventional weather analyses present recognizable patterns that precede and accompany the development of heavy snowstorms along the Northeast coast of the United States. These repeating patterns may indicate some of the physical processes necessary for heavy snowfall. Thus, these patterns provide a first step in the generation of conceptual models that can be used by forecasters, researchers, and students to predict future heavy snow events in a region so greatly affected by their occurrence.

The monograph also demonstrates that there is considerable case-to-case variability, supporting the notion that there are no foolproof scenarios that can be applied uniformly to all future storms. Every potential heavy snow situation needs to be examined carefully because no two weather systems evolve in exactly the same manner. Some cases will develop into heavy storms despite the fact that they do not match the conceptual models presented here. Conversely, a number of "false alarms" will occur in which the heavy snow does not materialize despite the appearance of favorable precursor conditions in accordance with the conceptual models.

A recognition of the limitations of this study based on only 20 cases over a 30-year period points to the need for additional research that involves detailed case studies, numerical modeling efforts, and continued climatological reviews. Detailed diagnostic studies of the processes associated with major East Coast storms, such as those produced by Uccellini et al. (1984, 1985, 1987), Bosart (1981), Bosart and Lin (1984), Sanders and Bosart (1985a,b), Bosart and Sanders (1986), and others, are needed to sort out the complex nonlinear interactions that produce major snowfalls. These diagnostic studies will increasingly rely on meso- to regional-scale numerical models to determine the storms' sensitivity to various physical processes, as discussed by Keyser and Uccellini (1987), and to determine the manner in which these processes interact to produce heavy snowfall along the Northeast coast. Numerical modeling studies could be performed on any of the cases presented here to identify the key processes, their important interactions, and how their effects vary from case to case.

A major issue that this monograph addresses but cannot resolve, and which appears to require the more detailed diagnostic and numerical modeling studies, is our need to understand the degree to which each physical process contributes to an extended period of heavy snow, rather than rain, along the coastal plain. While cyclones were observed in each case, many other cyclones during the winter season produce little or no snow along the Northeast coast. The evolution of the Canadian or polar anticyclones is a major factor that determines whether the precipitation falls as snow or as rain. The extent to which Canadian air masses can be warmed and moistened by their passage over the ocean and yet still be cold enough to sustain snow over the coastal plain needs to be determined. A study of the low-level airflow originating in the anticyclones would complement the existing models of airflow within midlatitude cyclones (e.g., Carlson 1980; Browning 1986; Young et al. 1987).

The physical interactions responsible for secondary cyclogenesis near the East Coast also need to be better understood. Secondary cyclogenesis, which occurs as a primary cyclone weakens near the Appalachian Mountains, is a common feature of many of the storms examined here. The timing and location of the secondary redevelopment are crucial, since the position of the heavy snowfall area is dependent upon the path of the secondary cyclone. Model-based diagnostic studies of many different storm systems are required to determine whether secondary cyclogenesis always develops from a similar combination of the physical processes listed in Table 11 or if various combinations produce the same result.

A related issue is the need to determine what factors are responsible for the enhanced deepening rates of surface cyclones over the Atlantic Ocean. Snowfall rates increase along the coastal plain of the Northeast during periods of rapid surface pressure falls. There does not appear to be a one-to-one correspondence between the deepening rates of the storms and the strength of the upper-level jet streaks, the amplitude of the upper-level trough/ridge systems, or the amount of precipitation. Therefore, questions concerning the roles of friction (e.g., Anthes and Keyser 1979), latent heat release (e.g., Danard 1964), sensible and latent heat fluxes (e.g., Bosart 1981), the interactions of upper- and lower-level potential vorticity maxima (e.g., Farrell 1984, 1985; Hoskins et al. 1985; Uccellini et al. 1985; Boyle and Bosart 1986; Whitaker et al. 1988), and convection (e.g., Tracton 1973; Bosart 1981; Gyakum 1983b) need to be addressed. In particular, it has been hypothesized that convection contributes to or even initiates cyclogenesis. Further evidence is needed to support this idea.

Another critical issue involves how upper-level trough/ridge systems and their associated jet streaks interact or "phase" to enhance or suppress cyclogenesis. This problem was demonstrated by the widely varying operational numerical model forecasts of the 22–23 February 1987 snowstorm (see section 8.3). Early numerical forecasts indicated that the interaction of two separate upper-level troughs would suppress the development of a cyclone along the East Coast. In reality, the two troughs combined to promote rapid cyclogenesis and snowfall rates of 2.5 to 12.5 cm h$^{-1}$ along the East Coast. Weather forecasters are frequently challenged by the difficulties in determining how multiple trough/jet systems will interact. Detailed theoretical, numerical, and case studies will be needed to assess these interactions.

The mesoscale structure of snow bands, as documented by Sanders and Bosart (1985a) and Sanders (1986b) and demonstrated by the narrow, intense snowband in the Washington, D.C., area during the "Veterans Day Snowstorm" of 11 November 1987, must also be investigated further. Concepts such as slantwise stability (Emanuel 1985; Sanders and Bosart 1985a; Seltzer et al. 1985; Sanders 1986b; Wolfsberg et al. 1986) could be tested in a wide variety of cases to determine their applicability to the banded snowfall distributions found in the cases presented here. The detection of lightning and thunder in many storms indicates the presence of convection and the possible role of gravity wave phenomena (Bosart and Sanders 1986). These factors could also have important influences on the snowfall rates and distributions, and are therefore worthy of further research.

Finally, we recognize that the cases reviewed in this monograph constitute a rather small sample, and that a continued effort is required to develop a larger group of cases covering a wider variety of storms. We envision that this 30-year record of major snowstorms will be extended forward in time as new storms occur; it could also be expanded further back in time to include all the major Northeast snowstorms for which data is available. For example, Weiss and Wagner (1987) have reconstructed surface weather maps for major snowstorms that have affected Philadelphia, PA, since 1888. The sample also needs to include storms that did not produce heavy snowfall, to help in distinguishing between the patterns that generate light rather than heavy snow and those that produce various precipitation types, or a changeover from one precipitation type to another.

It is only through a combination of research efforts ranging from climatological reviews to model-based case studies, together with improved observations of these systems, especially over the ocean, that our understanding of these storms will be enriched and our ability to forecast them with acceptable consistency may be realized.

PLATE 1. Park Avenue and 37th Street, New York City, 12 December 1960. Photo reprinted courtesy of the *New York Times*.

PLATE 2. Times Square, New York City, 7 February 1967. Photo reprinted courtesy of the *New York Times*/Neal Boenzi.

PLATE 2. Night scene in Boston, Massachusetts, 24 February 1969. Photo reprinted courtesy of the *Boston Globe*/James F. McDevitt.

PLATE 4. Hempstead, Long Island, New York, 6 February 1978. Photo reprinted courtesy of *Newsday*/K. Wiles Stabile.

PLATE 5. Route 128, south of Boston, Massachusetts, 8 February 1978. Photo reprinted courtesy of the *Boston Globe*/David Ryan.

PLATE 6.   Wisconsin Avenue, Georgetown, Washington, D.C., 19 February 1979. Copyright Fred Maroon.

PLATE 7.   New York City, 6 April 1982. Photo reprinted courtesy of the *New York Times*/Neal Boenzi.

PLATE 8. The Long Island Expressway, 11 February 1983. Photo reprinted courtesy of *Newsday*/J. Michael Dombroski.

# CHAPTER 7

# Analyses of 20 Major Snowstorms: 1955–1985

As noted throughout this monograph, the evolution of the surface and tropospheric conditions prior to and during the snowstorms cannot be generalized because of the case-to-case variability that characterizes these storms. Palmén and Newton (1969, p. 273) emphasize this problem by stating: "Cyclonic disturbances appear in such diverse forms that it is impossible to give a description that is uniformly applicable to all cases." Detailed examinations of one or two cases, although they can provide many important insights, have limited applicability in other situations with only superficial similarity. Problems with compositing antecedent conditions for East Coast storms have also been cited (Brandes and Spar 1971). Therefore, a set of conventional map analyses is presented in this chapter for each of the 20 storms to describe the unique aspects of each event as well as the similarities to the other cases.

The purpose of this chapter is to illustrate how these important weather systems organize and evolve prior to and during the development of heavy snowfall along the northeast coast of the United States. Each case study includes a total snowfall chart and 12-hourly sequences of surface weather maps, 850-mb charts and 500-mb geopotential height analyses with upper-level wind analyses superimposed. Infrared satellite image sequences are also displayed for the five most recent cases.

Each examination begins with a brief summary of the storm's major effects on the Northeast coastal region. The snowfall charts were constructed from daily snowfall measurements shown in *Climatological Data*. Accumulations in excess of 25, 50, and 75 cm were shaded with amounts for selected locations noted in the text.

The surface charts depict fronts, high- and low-pressure centers, isobars, and precipitation, with moderate to heavy precipitation shaded more heavily. The sequence of maps is timed so that the top right panel captured the sea-level low-pressure center along the East Coast, when most of the storms were deepening rapidly. A discussion of the surface charts for each case focuses on the anticyclone, cold air damming east of the Appalachians, coastal frontogenesis, and the intensification, track, and duration of the cyclone.

The evolution of the lower troposphere is described with 850-mb analyses. The 850-mb charts display geopotential height contours, isotherms, values of maximum 12-hourly geopotential height falls, and selected wind reports. The 850-mb low track and intensity, the evolution of the lower-tropospheric baroclinic zone, and the development of low-level wind maxima, or jet streaks, are highlighted in the accompanying discussions.

The upper troposphere is depicted by a map sequence which displays 500-mb geopotential height contours, the axes of upper-level jet streaks, isotachs of maximum winds derived from either 500-, 300-, or 200-mb NMC analyses (250-mb analyses were examined for the cases of April 1982 and February 1983), and the locations of cyclonic absolute vorticity maxima. The wind analyses may not necessarily reflect the maximum wind speeds associated with the jet streaks, which could lie between the standard pressure levels, but should present a consistent and representative view of multiple-level jet structure. These fields are examined to address how upper-level troughs and jet streaks evolve in space and time and how these features influence the surface cyclogenesis and associated precipitation. A general overview of the large-scale circulation patterns across the United States and Canada prior to cyclogenesis is presented to identify any recognizable patterns that precede or accompany the major snow events on the East Coast. Various aspects of the troughs and their flanking ridges, most notably the cyclonic vorticity and vorticity advection, amplitude, wavelength, trough axis orientation, and the development of a closed circulation at 500 mb, are highlighted in relation to the surface cyclogenesis and development of precipitation. Changes in the strength and location of upper-level wind maxima are also discussed. The shortening wavelength of the trough/ridge system and the increasing magnitudes of jet streaks, which have been discussed by Palmén and Newton (1969) and noted recently by Mullen (1983) and Uccellini et al. (1984) as being associated with nonlinear processes influential in the development of cyclones, are also reviewed. The associations between diffluent flow, negatively tilted troughs, and rapidly deepening cyclones are covered, since these factors point to the importance of along-stream variations in the upper-level height (mass) and wind fields in the development of the storm systems. The discussions of upper-level features also include mention of the trough and confluence zones over southeastern Canada and the northeastern United States, and their associated jet streaks.

Twelve-hourly GOES-East infrared imagery, corresponding closely to the times of the surface and upper-air analyses, are presented for each case since 1978 to depict the evolution of cloud systems during cyclogenesis.

Finally, reviews of three additional snowstorms that affected the Northeast coast during the 1986/1987 winter season are also included in Chapter 8. Because these systems display the same large degree of variability that characterized the previous 20 cases, they help to illustrate the many configurations these storms can take.


## 7.1   18–20 MARCH 1956

*General remarks*

- The first snowstorm of the 20-case sample did not affect a very large area, but the heaviest snow occurred across the densely populated sections of eastern Pennsylvania, New Jersey, southeastern New York, and southern New England. Local newspaper accounts indicate that this storm was poorly predicted, with local forecasts only calling for flurries the day the snow began in New York City. As a result, many travelers became stranded on the roads. An early account of the storm can be found in Mook and Norquest (1956).

This system followed a more widespread snowstorm on 15–16 March that produced similar snow amounts in the interior of New York and New England.

- Regions with snow accumulations exceeding 25 cm: southeastern Pennsylvania, New Jersey (except the extreme south), southeastern New York, Connecticut, Rhode Island, Massachusetts (except northwest), and southeastern New Hampshire.
- Regions with snow accumulations exceeding 50 cm: scattered areas of northern New Jersey, southeastern New York, Connecticut, and Massachusetts.
- Urban center snowfall amounts:

  | | |
  |---|---|
  | Washington, D.C.–National Airport | 1.7″ (4 cm) |
  | Baltimore, MD | 5.5″ (14 cm) |
  | Philadelphia, PA | 8.7″ (22 cm) |
  | New York, NY–Central Park | 13.5″ (34 cm) |
  | Boston, MA | 13.3″ (34 cm) |

- Other selected snowfall amounts:

  | | |
  |---|---|
  | Babylon, NY | 25.6″ (65 cm) |
  | Newark, NJ | 18.2″ (46 cm) |
  | Providence, RI | 14.7″ (37 cm) |
  | Hartford, CT | 14.0″ (36 cm) |
  | Trenton, NJ | 12.2″ (31 cm) |

*The surface chart*

- Cold air for the snowfall was provided by a weak, but intensifying, anticyclone well north of New York, which developed behind the major storm of 15–16 March. Sea-level pressures at the center of this anticyclone rose from 1021 mb at 0300 UTC 18 March to 1033 mb by 1500 UTC 19 March.
- Cold air damming was observed at 1530 UTC 18 March, as evidenced by an inverted sea-level pressure ridge from southeastern New York to northern Virginia. The damming occurred at the same time that heavier snowfall was beginning to develop over the Middle Atlantic states. No coastal frontogenesis was detected.
- The primary low-pressure center tracked southeastward from the northern Plains states to the Ohio Valley on 16–17 March in the wake of the intense storm that had advanced northeastward along the East Coast on 15–16 March. The cyclone was weak, as indicated by the small sea-level pressure gradients in the vicinity of the low center, and was associated with only scattered light precipitation.
- A secondary cyclone formed over Virginia during the early afternoon of 18 March and deepened slowly. The secondary development was not especially pronounced, even though the primary cyclone dissipated over West Virginia in only about 6 hours. Thundershowers formed across Virginia and North Carolina, however, and steady snowfall developed from Maryland to New York City, as the cyclone crossed the Virginia coast and moved over the Atlantic Ocean.
- Heavy snowfall commenced late on 18 March and continued into 19 March from Pennsylvania and New Jersey into southern New England, as the storm

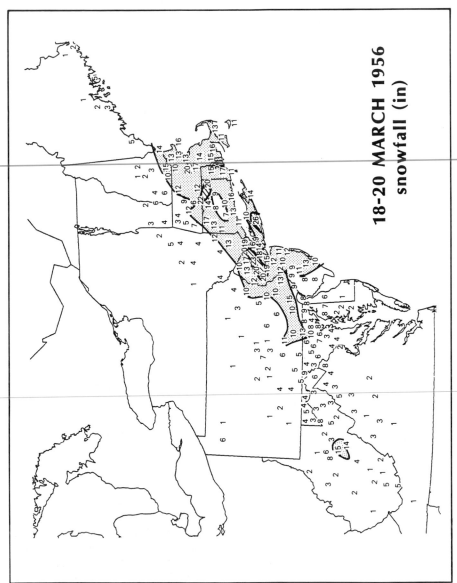

**18-20 MARCH 1956**
**snowfall (in)**

FIG. 30.    Snowfall (in) for 18–20 March 1956. Shading contours are for 10 in (25 cm), 20 in (50 cm), and 30 in (75 cm).

# SURFACE

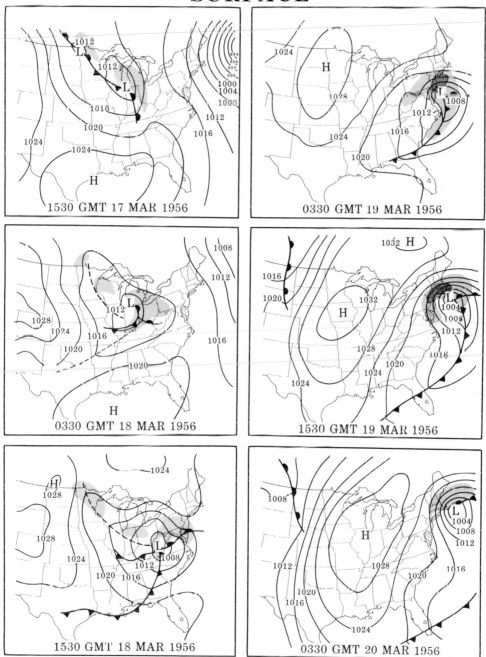

FIG. 31.   Twelve-hourly surface weather analyses for 1530 UTC 17 March 1956 through 0330 UTC 20 March 1956. Maps include surface high- and low-pressure centers and fronts; shading indicates precipitation (heavy shading denotes moderate to heavy precipitation), solid lines are isobars (4-mb intervals), and dashed lines represent axes of surface troughs not considered to be fronts.

drifted slowly northeastward and gradually deepened to 1000 mb. The central sea-level pressure dropped 8 mb over the 30-hour period ending at 1500 UTC 19 March.

- Northeasterly winds increased on 19 March as the sea-level pressure gradient north of the cyclone center increased.
- Snowfall gradually ended across southern New England late on 19 March as the cyclone continued drifting slowly east-northeastward at an average speed of 10 m s$^{-1}$.

*The 850-mb chart*

- Northwesterly flow dominated the northeastern United States prior to this storm, with the 0°C isotherm located over the southeastern United States on 17 March.
- The 850-mb low center, like its sea-level counterpart, did not intensify rapidly. The only deepening occurred east of the New Jersey coast between 0300 and 1500 UTC 19 March, when the geopotential height minimum fell 60 m. This 12-hour period of intensification occurred following the commencement of sea-level deepening.
- The region of maximum temperature gradient, including the 0°C isotherm, was initially displaced several hundred kilometers from the 850-mb low center on 17–18 March. The 0°C isotherm became nearly collocated with the cyclone center by 0300 UTC 19 March as the sea-level cyclone deepened east of Maryland.
- A pattern of enhanced 850-mb temperature advections was established by 0300 UTC 18 March, as the sea-level cyclone propagated through the Ohio Valley. At this time, strong cold air advection can be inferred over the eastern Plains states and the middle Mississippi Valley. By 1500 UTC 18 March, vigorous cold air advections west of the 850-mb low center and warm air advections east of the low were quite evident. An S-shaped isotherm pattern resulted near the East Coast late on 18 and early on 19 March.
- Southerly 850-mb winds over North Carolina and Virginia increased to 15 to 20 m s$^{-1}$ at 1500 UTC 18 March, as snowfall intensified across the Middle Atlantic states and convection developed in Virginia and North Carolina. By 0300 and 1500 UTC 19 March, a 20–25 m s$^{-1}$ LLJ was found north of the 850-mb low center. This easterly LLJ was collocated with the region of heavy snowfall, augmenting the moisture transport into this region.

*500-mb geopotential height analyses*

- The precyclogenetic environment over the United States and Canada before 19 March was dominated by a region of low geopotential heights over Quebec and eastern Canada, and a ridge over the extreme western United States and western Canada. A pronounced ribbon of large geopotential height gradients extended southeastward from western Canada through the Plains states to the southeastern United States.
- A confluent pattern can be discerned in the geopotential height field, especially

# 850 MB

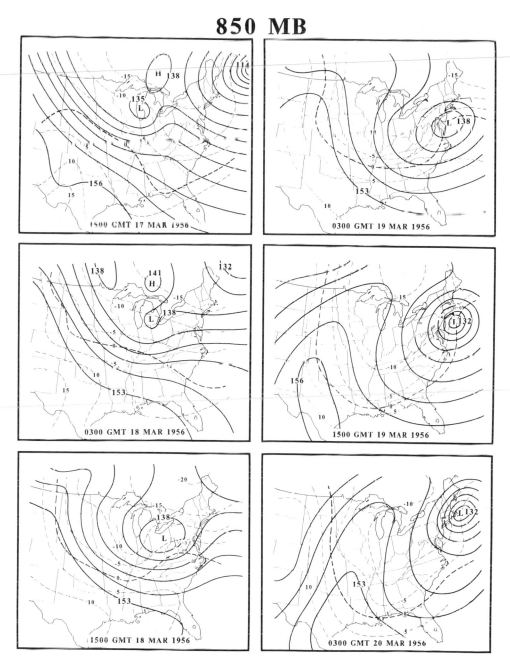

Fig. 32. Twelve-hourly 850-mb analyses for 1500 UTC 17 March 1956 through 0300 UTC 20 March 1956. Analyses include contours of geopotential height (solid, at 30 m intervals, 156 = 1560 m), isotherms (dashed, °C, 5°C intervals), selected values of geopotential height tendency [−3 = −30 m (12 h)$^{-1}$], and selected wind reports illustrating low-level wind maxima (flags represent 25 m s$^{-1}$, full barbs represent 5 m s$^{-1}$, and half barbs represent 2.5 m s$^{-1}$).

at 1500 UTC 18 March from the central Great Lakes and Midwest, eastward across West Virginia and New York. By 0300 UTC 19 March, this region had shifted to eastern New England eastward over the Atlantic. Rising sea-level pressures in association with an anticyclone over eastern Canada were located beneath the confluence.

- The 500-mb trough associated with the sea-level cyclone initially appeared as a slight perturbation in the northwesterly flow just north of North Dakota at 1500 UTC 17 March. The trough propagated southeastward and then eastward in the following 24 to 36 hours, reaching the Appalachian Mountains as the sea-level cyclone neared the East Coast at 0300 UTC 19 March. During this period, the axis of the trough rotated from a northeast–southwest tilt to a north–south orientation.

- A region of cyclonic absolute vorticity at 500 mb was associated with the southeastward-propagating trough. The vorticity maximum tracked from the Ohio Valley to the Middle Atlantic coast between 1500 UTC 18 March and 1500 UTC 19 March. The developing surface low and outbreak of heavy snow were located downwind of the propagating maximum in a region of cyclonic vorticity advection.

- By 1500 UTC 19 March, a center of low 500-mb geopotential heights (denoted by an "L") developed over New Jersey, as the trough appeared to rotate northeastward about a preexisting 500-mb low center near Lake Erie. In the following 12 hours, the new low center deepened 60 m and evolved into a separate vortex off the New England coast. The development of this vortex coincided with a period in which the sea-level cyclone did not appear to deepen, although it had deepened earlier.

- The 500-mb trough exhibited a large increase in amplitude, especially at 1500 UTC 18 March and 0300 UTC 19 March. The amplification occurred as heavy snow developed across the Middle Atlantic states and southern New England, and as the sea-level cyclone deepened, although at only a modest rate.

- The increase of amplitude was especially apparent because of the increase in 500-mb heights downwind of the trough axis, forming an amplifying ridge. The formation of the ridge at 1500 UTC 18 March and 0300 UTC 19 March occurred in conjunction with the development of strong warm air advection at 850 mb.

*Upper-level wind analyses*

- Through 1500 UTC 18 March, a jet streak at 300 mb extended across the northeastern United States, within confluent flow south of a trough over eastern Canada. The sea-level anticyclone, low-level northwesterly flow, and cold air advection were located beneath the entrance region of this jet streak, especially at 1500 UTC 18 March.

- Although gaps in wind coverage were common during the 1950s, two distinctly separate jet systems were observed across the southern United States throughout the course of this event.

- A subtropical jet was apparent at 200 mb across the southern United States,

# 500-MB HEIGHTS AND UPPER-LEVEL WINDS

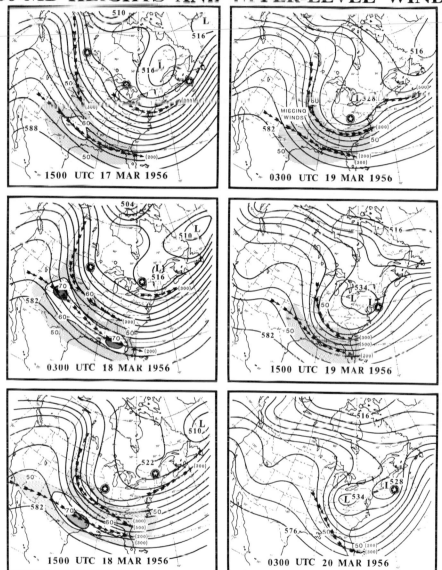

FIG. 33. Twelve-hourly analyses of 500-mb geopotential height and upper-level wind fields for 1500 UTC 17 March 1956 through 0300 UTC 20 March 1956. Analyses include locations of geopotential height maxima or minima, contours of geopotential height (solid, at 60-m intervals, 522 = 5220 m), locations of 500-mb absolute vorticity maxima (circled stars), axes of upper-level wind maxima (arrows, values in parentheses represent standard levels at which maximum velocities were observed), and isotachs of maximum wind speeds observed at either 500, 300, 250, or 200 mb (alternating intervals of shading represent 10 m s$^{-1}$ increments in wind speed).

with maximum velocities of 70 m s$^{-1}$ at 0300 and 1500 UTC 18 March, prior to cyclogenesis along the East Coast. Wind speeds within this jet increased between 1500 UTC 17 March and 0300 UTC 18 March, but did not appear to increase beyond this time (although there were many data omissions).

- A polar jet with maximum wind speeds of 50 to 60 m s$^{-1}$ was associated with the trough that spawned the cyclone. Maximum winds increased somewhat in the 24-hour period ending at 1500 UTC 18 March, as the sea-level cyclone approached the East Coast and intensified slightly. No further increases of wind speed were noted after this time. Secondary cyclogenesis occurred over Virginia in the diffluent exit region of the polar jet at 1500 UTC 18 March and 0300 UTC 19 March. The development of strong southerly winds at 850 mb at 1500 UTC 18 March was also coincident with the diffluent exit region of the polar jet.

## 7.2   14–17 FEBRUARY 1958

*General remarks*

- The "Blizzard of '58" was a very widespread storm that produced snow accumulations in excess of 25 cm from Alabama to Maine. The storm occurred during one of the stormiest and snowiest winters of the past half century. Intense cold and high winds persisted after the snow ended, prolonging the severe effects of the storm.
- Regions with snow accumulations exceeding 25 cm: central Virginia, central Maryland, northern Delaware, eastern Pennsylvania, New Jersey, New York (except southwest and northwest), Connecticut, Rhode Island, Massachusetts, Vermont, New Hampshire, and sections of western Maine, West Virginia, western Virginia, and Maryland.
- Regions with snow accumulations exceeding 50 cm: eastern Pennsylvania, western and eastern New York, southern Vermont, eastern Massachusetts, and scattered areas of New Hampshire, Connecticut, New Jersey, and Maryland.
- Urban center snowfall amounts:

| | |
|---|---|
| Washington, D.C.–National Airport | 14.4″ (37 cm) |
| Baltimore, MD | 15.5″ (39 cm) |
| Philadelphia, PA | 13.0″ (33 cm) |
| New York, NY–La Guardia Airport | 10.1″ (26 cm) |
| Boston, MA | 19.4″ (49 cm) |

- Other selected snowfall amounts:

| | |
|---|---|
| Callicoon, NY | 38.7″ (98 cm) |
| Rochester, NY | 30.6″ (78 cm) |
| Syracuse, NY | 26.5″ (67 cm) |
| Albany, NY | 18.3″ (46 cm) |
| New Haven, CT | 17.2″ (44 cm) |
| Allentown, PA | 15.8″ (40 cm) |

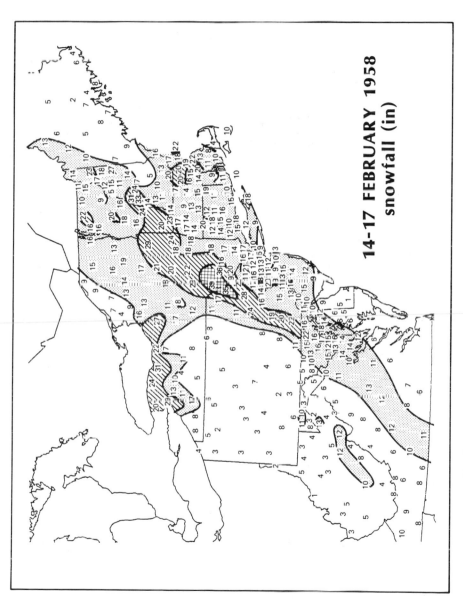

**14-17 FEBRUARY 1958 snowfall (in)**

FIG. 34.   Snowfall (in) for 14-17 February 1958. See Fig. 30 for details.

Worcester, MA                                               13.6″ (35 cm)
Newark, NJ                                                  13.3″ (34 cm)
Trenton, NJ                                                 13.0″ (33 cm)

*The surface chart*

- As one anticyclone moved off the East Coast on 14 February, cold air for the snowfall was reinforced by another large anticyclone (1035 mb) moving southward from the Canadian Plains toward the northern Plains states. A lobe of high pressure extended eastward from this anticyclone across Ontario and into southern Quebec.

- Cold air damming was observed over the southeastern United States on 15 February, as an inverted ridge of sea-level pressure developed from North Carolina to Georgia. Coastal frontogenesis occurred immediately off the Southeast coast at the same time and coincided with the development of moderate to heavy precipitation across Georgia and the Carolinas.

- The sea-level cyclone followed a path from the western Gulf of Mexico to Georgia, then northeastward up the Atlantic coast. The cyclone center appeared to jump or redevelop northeastward when it reached South Carolina at 1800 UTC 15 February. A new center formed over eastern North Carolina and reached southeastern Virginia by 0000 UTC 16 February. The average propagation rate of the storm was 17 m s$^{-1}$, but during this 6-hour period, the rate increased to nearly 25 m s$^{-1}$. The forward rate of motion slowed to about 10 m s$^{-1}$ as the cyclone neared southern New England on 16 February.

- A weak, inverted sea-level pressure trough extending northward from the cyclone center during 14–15 February was accompanied by light snowfall. This feature reflected an upper-level trough and vorticity maximum that was separate from the upper-level system associated with the cyclone. The inverted sea-level trough lost its identity late on 15 February as it became incorporated into the expanding circulation about the deepening cyclone.

- Sea-level deepening occurred nearly continuously from 14 February to 16 February as the cyclone moved from the western Gulf of Mexico to the New England coast. During this period, the central pressure fell from 1004 to 970 mb. Deepening rates exceeding $-3$ mb $(3\ h)^{-1}$ prevailed between 1200 UTC 15 February and 1200 UTC 16 February. Heavy precipitation was observed along the East Coast during this period of rapid intensification.

- Concurrent with the rapid deepening of the sea-level cyclone, strong northeasterly winds developed north of the storm center along the Middle Atlantic and southern New England coasts. As the cyclone moved off the New England coast on 16–17 February, bitterly cold air followed in its wake. Temperatures fell from $-10°C$ to $-20°C$ across the northeastern United States on 17–18 February.

*The 850-mb chart*

- As the cyclone was developing over Texas and Louisiana on 14 February, cold air dominated the lower troposphere over the central and eastern United States, with the 0°C isotherm suppressed over the southern United States. At

# SURFACE

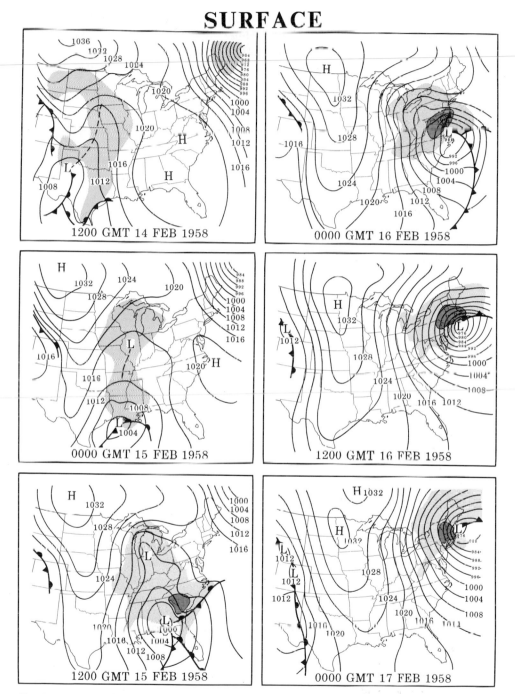

FIG. 35.   Twelve-hourly surface weather analyses for 1200 UTC 14 February 1958 through 0000 UTC 17 February 1958. See Fig. 31 for details.

the same time, strong northwesterly flow behind an intense cyclone east of Maine was associated with 850-mb temperatures between $-10°$ and $-20°C$ in the northeastern states. The northwesterly flow was located beneath a confluent upper-level geopotential height pattern over the northeastern United States.

- The 850-mb low center formed over Louisiana by 0000 UTC 15 February and deepened during the next 48 hours. Maximum deepening occurred as the center moved from Alabama to southeastern Virginia $[-90 \text{ m } (12 \text{ h})^{-1}]$ and then to near Long Island $[-150 \text{ m } (12 \text{ h})^{-1}]$ between 1200 UTC 15 February and 1200 UTC 16 February. This period coincided with the rapid intensification of the sea-level cyclone. The 850-mb center then deepened only 30 m as it moved to near Boston, Massachusetts, by 0000 UTC 17 February, an indication of the occlusion of the cyclone.

- The 850-mb low evolved along a band of isotherms over the southern United States, where the temperature gradient increased between 14 and 15 February. The development of the low center at 0000 UTC 15 February was accompanied by a strong surge of cold air advection west of the low over Texas. In the next 12 hours, this cold air surge was followed by the development of a pronounced southerly LLJ over the southeastern United States. The development of this jet coincided with an outbreak of moderate to heavy precipitation over the southeastern states and the formation of the coastal front.

- The enhanced cold air advection west of the low center and warm air advection east of the low center created an S-shaped isotherm pattern over the eastern United States on 15 February.

- The 0°C isotherm was nearly collocated with the 850-mb low center until the cyclone reached New England at 0000 UTC 17 February, at which time the center was located in colder air (another indication of occlusion). The 0°C isotherm appeared to represent the rain–snow demarcation rather well for this case, since it progressed no further northward than the Delaware/Maryland coast (compare with snowfall chart).

- By 1200 UTC 16 February, an easterly wind maximum was established north of the 850-mb low center as it deepened rapidly, and was located over regions experiencing heavy snowfall.

*500-mb geopotential height analyses*

- The precyclogenetic period, prior to 15 February, featured an upper-level ridge across the western United States and Canada, an intense cyclonic circulation over extreme southeastern Canada, a separate cyclonic vortex near Hudson Bay, and a closed anticyclonic circulation south of Greenland. The greatest geopotential height gradients were located over the United States, where several jet streams were observed. A trough was propagating from the southwestern United States into the south-central states.

- The surface high-pressure ridge and cold temperatures in the northeastern United States were located beneath a distinct region of confluent geopotential heights upwind of a deep 500-mb trough over southeastern Canada on 14–15 February.

# 850 MB

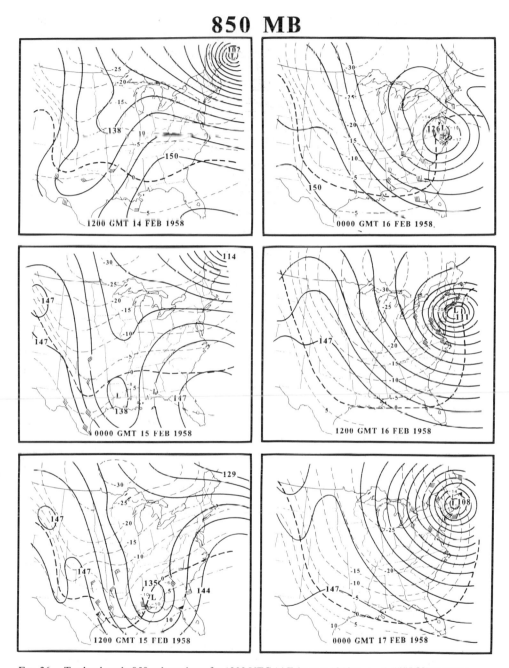

FIG. 36.  Twelve-hourly 850-mb analyses for 1200 UTC 14 February 1958 through 0000 UTC 17 February 1958. See Fig. 32 for details.

- The intensifying sea-level cyclone was associated with an upper-level trough that tracked eastward across the southern United States and then rotated northeastward along the East Coast on 16 February.
- A separate upper-level trough associated with the sea-level inverted trough over the Ohio Valley rotated about the Hudson Bay vortex between 1200 UTC 14 February and 1200 UTC 15 February. It then appeared to merge or phase with the trough moving northeastward along the East Coast at 0000 UTC 16 February.
- The deepening cyclone and heavy precipitation were associated with cyclonic vorticity advection downwind of a vorticity maximum crossing the southern and eastern United States on 14–16 February. The sea-level inverted trough over the Ohio Valley on 14–15 February was associated with a separate vorticity maximum that probably merged with the other vorticity field by 0000 UTC 16 February.
- As the trough associated with the cyclone moved northeastward along the Atlantic seaboard and merged with the other trough swinging around the vortex near Hudson Bay, geopotential heights fell rapidly along the East Coast between 1200 UTC 15 February and 0000 UTC 17 February. During this period, the sea-level cyclone deepened rapidly and a major shift in the upper-level circulation pattern occurred, as the Hudson Bay vortex became reestablished over the northeastern United States.
- The sea-level cyclone underwent its greatest intensification after 1200 UTC 15 February within the diffluent region downwind of the upper-level trough as the trough axis neared the East Coast and changed its orientation from north–south to northwest–southeast (a negative tilt), and as the 500-mb low center either propagated, or more likely redeveloped over the northeastern United States.
- Increases in the amplitude of the trough and the downstream ridge during cyclogenesis were small relative to the other cases (see Table 7). The half-wavelength between the trough and downstream ridge decreased substantially, especially between 1200 UTC 15 February and 1200 UTC 16 February.

*Upper-level wind analyses*

- Serious gaps in wind observation coverage plagued this case, especially over the western United States, but the event featured a variety of jet systems that influenced the development of the cyclone.
- A 50 m s$^{-1}$ polar jet was in evidence at 1200 UTC 14 February and 0000 UTC 15 February across the northeastern United States to the south of an intense vortex at 500 mb over extreme southeastern Canada. The cold air outbreak across the eastern United States occurred within the confluent entrance region of this jet.
- A subtropical jet with velocities exceeding 80 m s$^{-1}$ was observed over the southeastern United States through 1200 UTC 15 February, when its axis was located south of the developing surface low.
- By 1200 UTC 15 February, a separate 60 m s$^{-1}$ jet had developed from the

# 500-MB HEIGHTS AND UPPER-LEVEL WINDS

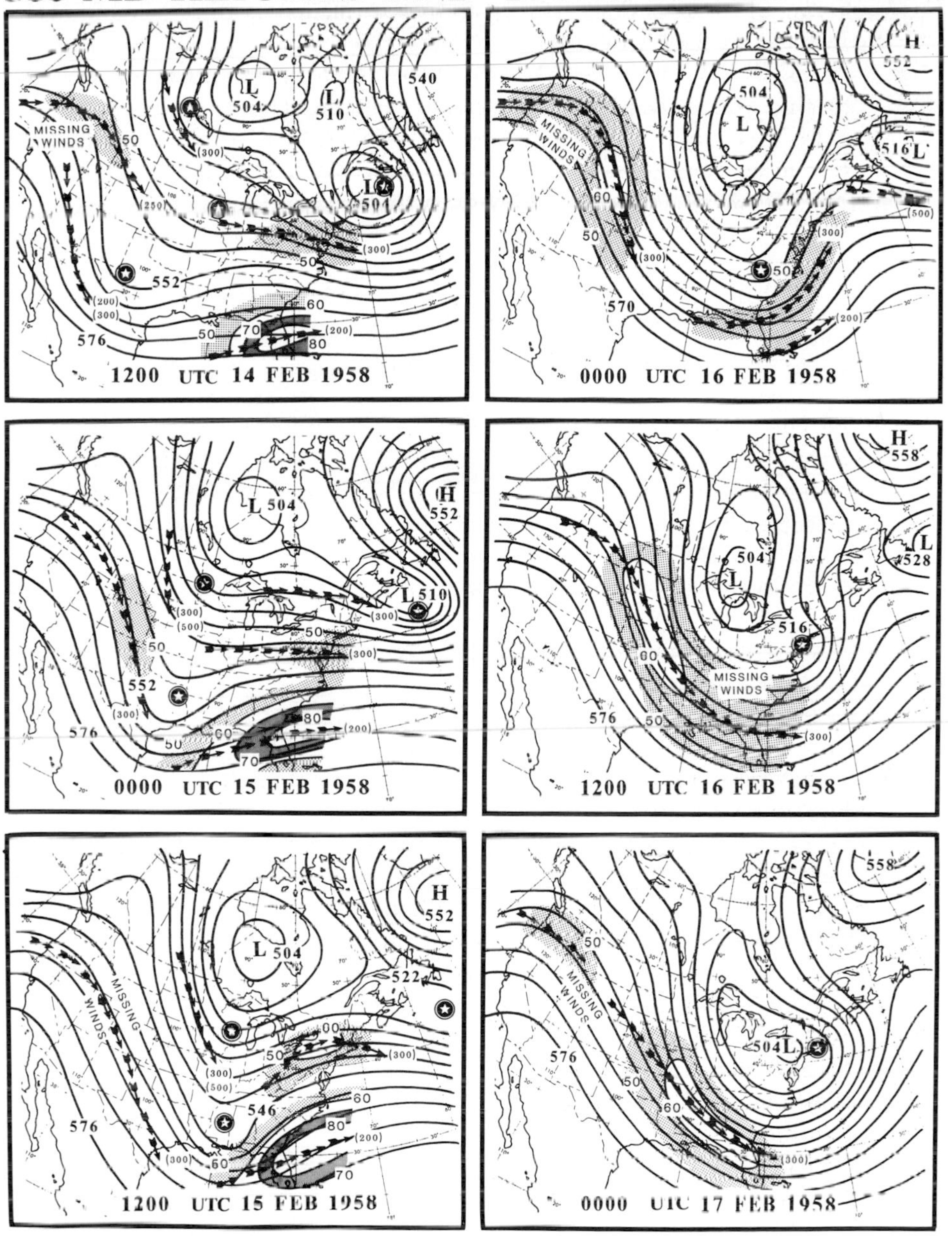

FIG. 37.    Twelve-hourly analyses of 500-mb geopotential heights and upper-level wind fields for 1200 UTC 14 February 1958 through 0000 UTC 17 February 1958. See Fig. 33 for details.

upper Ohio Valley to the northeastern United States at the crest of an amplifying upper-level ridge. This jet formed immediately prior to rapid cyclogenesis along the East Coast.

- A polar jet that was difficult to analyze because of missing wind reports was observed west of the upper trough propagating through the south-central states on 14–15 February. Rapid cyclogenesis appeared to commence within the exit region of this jet as the jet reached the base of the deepening upper-level trough between 1200 UTC 15 February and 0000 UTC 16 February.

## 7.3   18–21 MARCH 1958

*General remarks*

- This late-winter snowstorm was one of the more unusual cases of the sample. Elevation played a very significant role in the snowfall distribution since this was the *warmest* storm of the 20 cases. Although snow amounts at the official reporting sites in the five largest cities did not reach 30 cm, some sections of Washington, D.C., Baltimore, Philadelphia, and New York City received from 35 to more than 50 cm. While many local forecasters from Washington to Boston assumed that precipitation would fall primarily as rain, heavy, wet snow with temperatures near or slightly above freezing resulted in tremendous destruction of power lines and foliage. In parts of eastern Pennsylvania and northern Maryland, this was the greatest snowfall on record. Further descriptions of the storm can be found in Sanderson and Mason, Jr. (1958).
- Regions with snow accumulations exceeding 25 cm: northern Virginia, central Maryland, northern Delaware, eastern Pennsylvania, New Jersey (except near the coast), southeastern New York, western Connecticut, and scattered areas of eastern West Virginia, western Pennsylvania, Rhode Island, east-central Massachusetts, New Hampshire, and Maine.
- Regions with snow accumulations exceeding 50 cm: sections of north-central Maryland, eastern Pennsylvania, western New Jersey, and scattered areas of northern Virginia, northern Delaware, southeastern New York, Massachusetts, and New Hampshire.
- Urban center snowfall amounts:

| | |
|---|---|
| Washington, D.C.–National Airport | 4.8″ (12 cm) |
| Baltimore, MD | 8.4″ (21 cm) |
| Philadelphia, PA | 11.4″ (29 cm) |
| New York, NY–Central Park | 11.7″ (30 cm) |
| Boston, MA | 6.7″ (17 cm) |

- Other selected snowfall amounts:

| | |
|---|---|
| Morgantown, PA | 50.0″ (127 cm) |
| Mt. Airy, MD | 33.0″ (84 cm) |
| Allentown, PA | 20.3″ (51 cm) |
| Wilmington, DE | 19.0″ (48 cm) |
| Worcester, MA | 18.8″ (48 cm) |

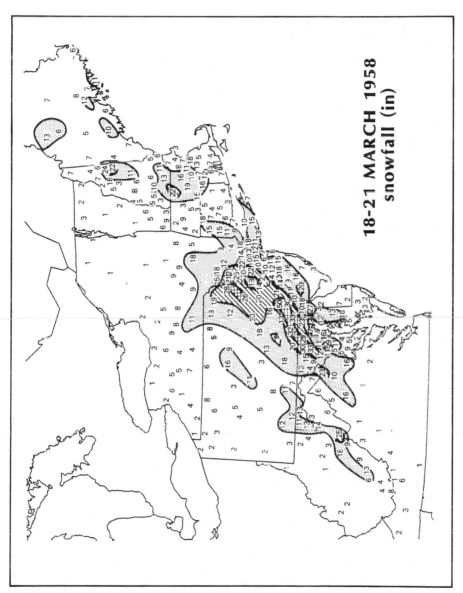

FIG. 38.   Snowfall (in) for 18–21 March 1958. See Fig. 30 for details.

| Trenton, NJ | 17.8″ (45 cm) |
| Newark, NJ | 14.8″ (38 cm) |

*The surface chart*

- Cold air (or, more appropriately, cool air) for this event was provided by a large anticyclone centered over extreme northern Canada, with a weak ridge of high pressure extending south-southeastward into the northeastern United States. On 19 March, late-winter sunshine helped to boost surface temperatures in this air mass to as high as 10°C in Maine. Where clouds and precipitation dominated, however, temperatures generally ranged between 0° and 2°C.

- Weak cold air damming was observed, especially at 0000 UTC 19 March, when an inverted high-pressure ridge extended from New Jersey to South Carolina. At the same time, coastal frontogenesis, which was more pronounced in the wind field than in the temperature field, was occurring off the Carolina coast.

- The most striking characteristic of the sea-level cyclone was its slow movement, which averaged only 6 m s$^{-1}$ during the 60-hour period shown in the analyses. The cyclone headed northeastward along the Southeast coast at a rate of approximately 10 m s$^{-1}$ on 19 March, but then slowed to 5 m s$^{-1}$ or less on 20–21 March off the New Jersey and southern New England coasts. The low-pressure center then drifted ashore over Rhode Island on 21 March. The slow movement of the cyclone resulted in a three-day period of steady precipitation, with melted totals greater than 10 cm over portions of Maryland and Pennsylvania.

- No redevelopment occurred in this case, at least during the period studied.

- The cyclone was initially rather weak, as indicated by the small pressure gradients surrounding the low center off the Southeast coast on 19 March. It deepened continuously through 0000 UTC 21 March, becoming a rather intense storm on 20 March off the Middle Atlantic coast. The greatest deepening occurred between 0000 UTC 20 March and 0000 UTC 21 March, when the central pressure fell from 997 to 976 mb.

- The heaviest precipitation occurred from late on 19 March through early on 21 March, as the cyclone underwent its greatest intensification. Strong northeasterly winds developed from the Middle Atlantic coast to the southern New England coast during the same period.

*The 850-mb chart*

- A weak 850-mb anticyclone was located over New England just prior to the storm. Northwesterly flow had prevailed over the northeastern United States on 17–18 March, but was confined to southeastern Canada by 19 March.

- While the 850-mb low center was located on the Alabama–Georgia border at 0000 UTC 19 March, the remnants of a weakening short-wave trough nearing the East Coast brought a separate region of relatively low geopotential heights, large temperature gradients, and 10 m s$^{-1}$ southeasterly winds to the Carolina coast. These features remained near the Carolina coastline at 1200 UTC 19 March and were related to the light precipitation occurring along the East Coast during this period.

# SURFACE

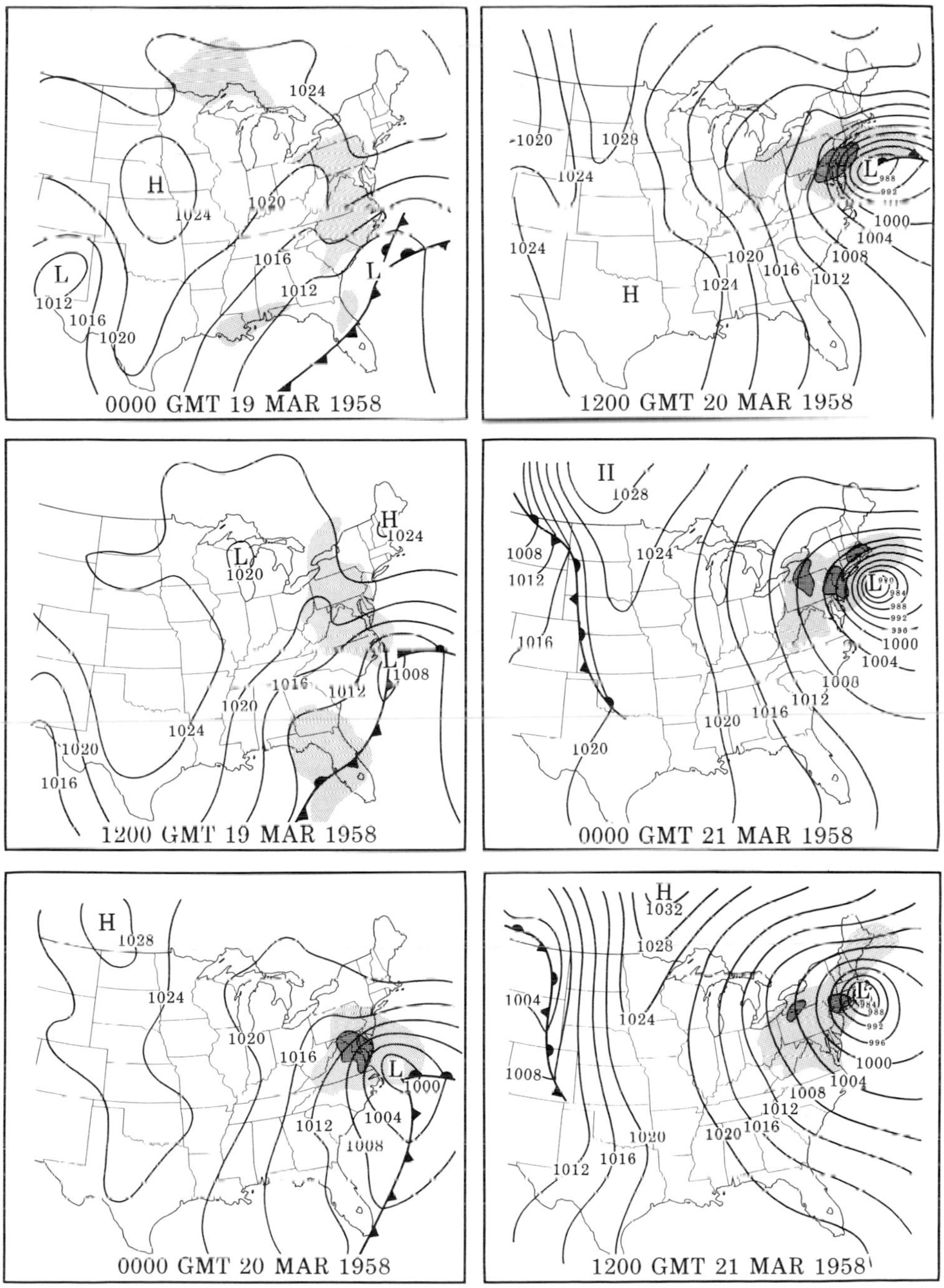

FIG. 39.   Twelve-hourly surface weather analyses for 0000 UTC 19 March 1958 through 1200 UTC 21 March 1958. See Fig. 31 for details.

# 850 MB

FIG. 40.　Twelve-hourly 850-mb analyses for 0000 UTC 19 March 1958 through 1200 UTC 21 March 1958. See Fig. 32 for details.

- By 0000 UTC 20 March, a well-defined and strengthening 850-mb center was in evidence over northeastern North Carolina as the sea-level cyclone was beginning to intensify offshore. This center deepened rapidly during the following 24 hours as it moved to near Long Island.
- The 850-mb low developed along a preexisting band of isotherms which extended from the Gulf of Mexico to the North Carolina and Virginia coasts on 19 March. The low center was located slightly to the warm side of the 0°C isotherm through 1200 UTC 20 March, then shifted to the cold side while the cyclone was occluding on 21 March. The 0°C isotherm roughly coincided with the demarcation between heavy and light snowfall amounts.
- Strong easterly winds developed north of the low center during the period of rapid intensification, with 35 m s$^{-1}$ speeds over New York City at 1200 UTC 20 March and 0000 UTC 21 March. The development of heavy precipitation at these times suggests that the easterly LLJ increased the moisture transport into the region of heavy precipitation.

*500-mb geopotential height analyses*

- The precyclogenetic environment was characterized by a trough that bisected the central United States, with a pronounced ridge over Hudson Bay and a deep cyclonic vortex east of Labrador. The ridge and vortex over eastern Canada remained relatively stationary during the 60-hour study period.
- The surface anticyclone over northern Canada and its associated high-pressure ridge extending southeastward into New England on 19–20 March were associated with the stationary 500-mb anticyclone near Hudson Bay and the 500-mb cyclonic vortex over extreme eastern Canada.
- A 500-mb trough of small amplitude and short wavelength weakened as it reached the East Coast. By 0000 UTC 19 March, this feature was nearly unidentifiable in the height field, but was still associated with a weak sea-level cyclone off South Carolina and light precipitation along the East Coast.
- A broad trough with no sharply defined axis was located over the central and eastern half of the United States at 0000 UTC 19 March. During the following 48 hours, this trough amplified slowly and evolved into a large closed cyclonic circulation over the northeastern United States. The vortex first developed over the upper Midwest at 1200 UTC 19 March, then redeveloped near the East Coast by 0000 UTC 21 March as the sea-level cyclone deepened rapidly. The vortex intensified considerably by 1200 UTC 21 March, as it became nearly vertical with the surface system.
- As the broad trough amplified on 19–20 March, several poorly defined, shorter-wavelength trough features rotated about the 500-mb low center located over the upper Midwest. While these features were difficult to distinguish in the geopotential height field, they were revealed by the height tendencies and were associated with distinct cyclonic vorticity maxima. Two of these short-wave features propagated toward the East Coast during 1200 UTC 19 and 20 March before merging near Long Island at 0000 UTC 21 March. The rapid development of the sea-level cyclone near the East Coast on 20 March occurred

# 500 MB HEIGHTS AND UPPER-LEVEL WINDS

FIG. 41.   Twelve-hourly analyses of 500-mb geopotential heights and upper-level wind fields for 0000 UTC 19 March 1958 through 1200 UTC 21 March 1958. See Fig. 33 for details.

    downwind of these propagating features, in a region of cyclonic vorticity advection.

- Rapid sea-level deepening commenced by 0000 UTC 20 March as a region of upper-level diffluence, which had developed downwind of the trough axis, crossed the coastline.

- The half-wavelength between the trough axis and the downstream ridge decreased rapidly during cyclogenesis, especially between 1200 UTC 19 March and 1200 UTC 20 March.

*Upper-level wind analyses*

- Two separate jet systems were located along the East Coast prior to the rapid surface development. A subtropical jet, whose axis crossed Florida, appeared to either weaken or move off the coast when maximum wind speeds dropped from 70 to 50 m s$^{-1}$ between 0000 UTC 19 March and 0000 UTC 20 March. Meanwhile, a polar jet streak across the Middle Atlantic states contained maximum wind speeds of 50 m s$^{-1}$ near 300 mb.
- Rapid deepening of the cyclone, which had begun by 0000 UTC 20 March, occurred between the polar and subtropical jets. The polar jet had shifted northward by this time, and maximum speeds had increased to 60 m s$^{-1}$ at the crest of the ridge near the New England coast. The subtropical jet remained over the Southeast coast.
- One or two jet streaks were observed upwind of the trough axis as it neared the East Coast and amplified between 0000 UTC 19 March and 1200 UTC 20 March. These jets may have propagated beneath the subtropical jet over the southeastern United States, contributing to an increase in the upper-level winds and associated cyclonic wind shears, thereby focusing the vorticity field into one distinct maximum near Long Island by 0000 UTC 21 March.

## 7.4   2–5 MARCH 1960

*General remarks*

- This widespread late-winter storm was especially fierce in eastern New England. Severe blizzard conditions occurred in eastern Massachusetts, as snow accumulations exceeded 50 to 75 cm and near-hurricane force winds battered the coast.
- Regions with snow accumulations exceeding 25 cm: eastern West Virginia, northern and western Virginia, much of Pennsylvania, parts of Maryland and Delaware, southeastern and western New York, Connecticut, Rhode Island, Massachusetts, southern Vermont, New Hampshire, and Maine.
- Regions with snow accumulations exceeding 50 cm: eastern Massachusetts, Rhode Island, and scattered areas of northern New Jersey, southeastern New York, Connecticut, and New Hampshire.
- Urban center snowfall amounts:

| | |
|---|---|
| Washington, D.C.–National Airport | 7.9″ (20 cm) |
| Baltimore, MD | 10.4″ (26 cm) |
| Philadelphia, PA | 8.4″ (21 cm) |
| New York, NY–La Guardia Airport | 15.5″ (39 cm) |
| Boston, MA | 19.8″ (50 cm) |

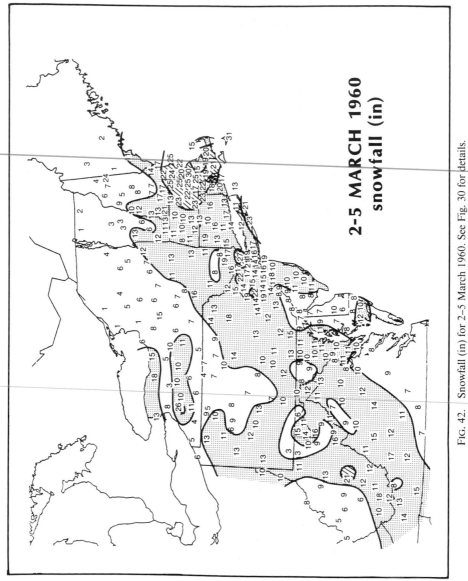

Fig. 42.  Snowfall (in) for 2–5 March 1960. See Fig. 30 for details.

- Other selected snowfall amounts:

| | |
|---|---|
| Nantucket, MA | 31.3" (79 cm) |
| Blue Hill Observatory, Milton, MA | 30.3" (77 cm) |
| Worcester, MA | 22.1" (56 cm) |
| Scranton, PA | 18.0" (46 cm) |
| Providence, RI | 17.7" (45 cm) |
| Roanoke, VA | 17.4" (44 cm) |
| Allentown, PA | 14 4" (37 cm) |
| Hartford, CT | 13.0" (33 cm) |
| Pittsburgh, PA | 12.8" (33 cm) |

*The surface chart*

- A large, cold anticyclone preceded the storm. The axis of highest sea-level pressure extended from the Great Lakes states to the Middle and South Atlantic coasts on 2–3 March. A pronounced inverted high-pressure ridge along the South Atlantic coast indicated cold air damming, especially on 2 March. Coastal frontogenesis occurred off the Southeast coast on 2 March, prior to the development of a secondary sea-level cyclone.
- The primary low-pressure center moved from the Gulf Coast to the Ohio Valley on 2–3 March. Moderate to heavy precipitation fell across the Tennessee and Ohio valleys in association with this system. The low deepened 8 mb in the 12 hours ending at 0000 UTC 3 March, prior to the development of the secondary low, then deepened slowly and erratically before finally dissipating during 3 March. The primary low was evident for 15 hours after the onset of secondary cyclogenesis.
- The secondary low-pressure center formed near the South Carolina coast between 0000 and 0600 UTC 3 March. The development occurred along a pronounced coastal front off the Southeast coast, as cold air wedged as far south as northern Florida.
- The secondary low center deepened 45 mb in the 30-hour period from 0600 UTC 3 March to 1200 UTC 4 March, with the central pressure falling to 960 mb off New England. Heavy snowfall and high winds were widespread across the Middle Atlantic states and southern New England during this phase of rapid development.
- The movement of the storm center from the Carolinas to off the New Jersey coast on 3 March was quite rapid, averaging greater than 20 m s$^{-1}$. It subsequently slowed to less than 10 m s$^{-1}$ off the New England coast on 4 March, as the lowest pressure was reached. The deceleration of the cyclone near New England was attended by strong easterly flow, prolonging the snowfall and enhancing accumulations across southeastern New England.

*The 850-mb chart*

- Northwesterly flow maintained 850-mb temperatures less than −10°C north of Virginia through 1200 UTC 2 March.
- The 850-mb low center propagated east-northeastward from the southern Plains states to the East Coast on 2–4 March. Its deepening rate increased from −60 m (12 h)$^{-1}$ in the 12-hour period ending at 0000 UTC 3 March to

# SURFACE

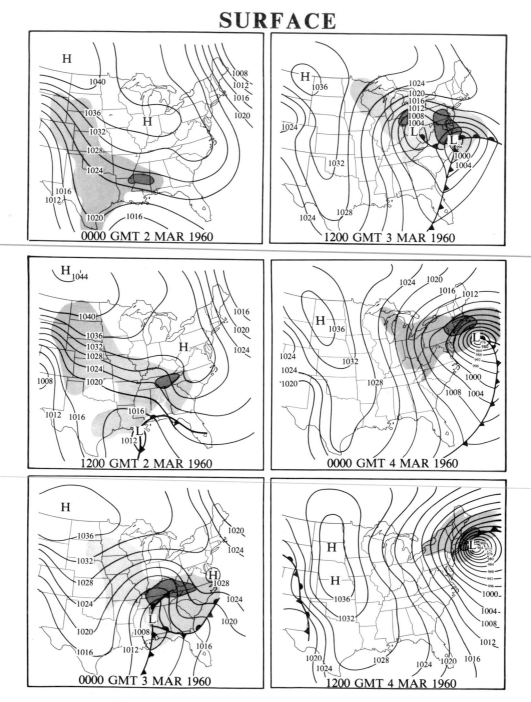

FIG. 43.   Twelve-hourly surface weather analyses for 0000 UTC 2 March 1960 through 1200 UTC 4 March 1960. See Fig. 31 for details.

$-90$ m $(12$ h$)^{-1}$ by both 1200 UTC 3 March and 0000 UTC 4 March, and to $-150$ m $(12$ h$)^{-1}$ by 1200 UTC 4 March. The increase in the 850-mb deepening rate coincided with the rapid intensification of the secondary surface low after 0000 UTC 3 March.

- The elongation of the 850-mb low at 1200 UTC 3 March suggests two centers, with the southeastern extension over Virginia reflecting the rapid development of the secondary surface low.

- A concentrated isotherm ribbon moved from the southern United States on 2 March northward to the Middle Atlantic states by 3 March. The thermal gradient intensified by 1200 UTC 3 March, as the secondary surface low developed. The 0°C isotherm was generally collocated with the 850-mb low center at 0000 UTC and 1200 UTC 3 March.

- Strong cold and warm air advection accompanied the 850-mb low center at 0000 UTC 3 March, producing an S-shaped isotherm pattern over the eastern United States. The development of this pattern coincided with the onset of rapid deepening of both the 850-mb and surface lows. The 850-mb low was located just west of the apex of the thermal ridge.

- A southerly LLJ was observed over Texas at 0000 UTC 2 March, over Alabama at 1200 UTC 2 March, and from Alabama to South Carolina by 0000 UTC 3 March, prior to rapid secondary development. The jet was associated with strong warm air advection across the southeastern United States as the primary surface low and 850-mb center deepened slowly. Winds over the southeastern United States increased from 15 to 20 m s$^{-1}$ to greater than 25 m s$^{-1}$ by 0000 UTC 3 March. Moderate to heavy precipitation broke out from Tennessee to southwestern Virginia, immediately downwind of this jet (see corresponding surface charts).

- Heavy snow developed along the Middle Atlantic coast by 1200 UTC 3 March as the secondary low formed, local geopotential height falls intensified, the temperature gradient increased, and a low-level east-to-southeasterly jet formed north of the 850-mb low center. The enhanced temperature gradient combined with increasing low-level winds to intensify the warm air advection and moisture transport into the developing precipitation area covering the Middle Atlantic states.

*500-mb geopotential height analyses*

- Strong geopotential height gradients and westerly flow dominated the precyclogenetic period across the United States into 2 March. Weak geopotential height gradients and large, nearly stationary vortices were observed over Canada, including a closed anticyclonic circulation between Quebec and Greenland, and cyclonic circulations over central and southeastern Canada.

- A 500-mb trough exiting southeastern Canada on 2 March produced confluent flow across the northeastern United States. This pattern was associated with strong northwesterly flow and cold air advection at low levels.

- The 500-mb trough associated with the sea-level cyclone extended from the United States–Canada border into Mexico at 0000 UTC 2 March, then propagated to the East Coast by the end of 3 March. A closed 500-mb center

# 850 MB

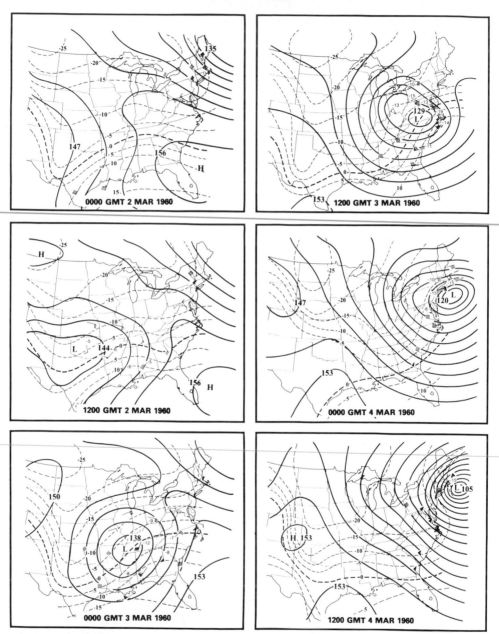

FIG. 44.   Twelve-hourly 850-mb analyses for 0000 UTC 2 March 1960 through 1200 UTC 4 March 1960. See Fig. 32 for details.

formed during rapid cyclogenesis late on 3 March. It then deepened rapidly on 4 March, as the surface low moved slowly off the New England coast.

- The trough axis became oriented from northwest to southeast (a negative tilt) during 2 March, as the primary cyclone deepened. The geopotential height gradient at the base of the trough increased as the secondary low-pressure

# 500-MB HEIGHTS AND UPPER-LEVEL WINDS

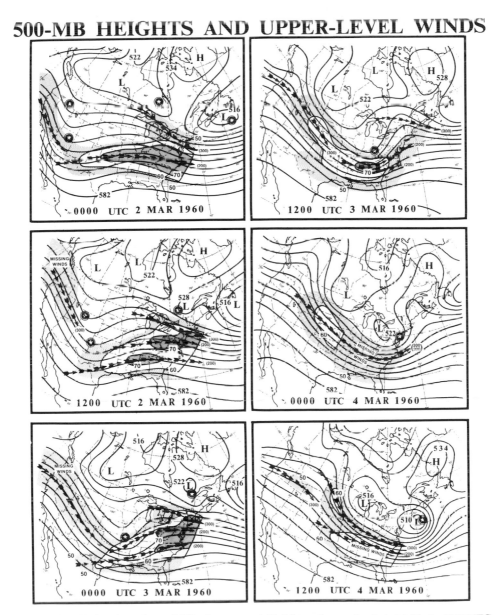

FIG. 45. Twelve-hourly analyses of 500-mb geopotential heights and upper-level wind fields for 0000 UTC 2 March 1960 through 1200 UTC 4 March 1960. See Fig. 33 for details.

center developed and intensified along the East Coast on 3–4 March. This secondary cyclogenesis occurred beneath a pronounced diffluence region downwind of the 500-mb trough.

- The cyclone and heavy snowfall developed downwind of a propagating and intensifying cyclonic vorticity maximum, in the region of strong vorticity advection.

- Although the surface cyclonic development was explosive in this case, changes in the amplitude of the 500-mb trough and downstream ridge were relatively small. There was some increase in amplitude of the trough and downstream ridge over the northern United States during secondary cyclogenesis on 3 March. Little increase in the amplitude of the trough and upstream ridge was observed prior to cyclogenesis.

- Although changes in amplitude were small, decreases in the half-wavelength between the trough and downstream ridge were substantial. They appeared to occur by 1200 UTC 2 March, during primary cyclogenesis, and by 0000 UTC 4 March, during secondary cyclogenesis, an apparent consequence of the "self-development" processes described earlier in the text.

*Upper-level wind analyses*

- A 300-mb jet crossed the northeastern United States within the confluent region upstream of the trough axis over southeastern Canada through 0000 UTC 3 March. The confluent entrance region of this polar jet was located above the cold high-pressure ridge that became entrenched over the northern United States by 0000 UTC 3 March.

- The jet stream crossed the entire country on 2–3 March within the zone of large geopotential height gradients that spanned the United States.

- An intense 200-mb subtropical jet streak, with maximum wind speeds in excess of 70 m s$^{-1}$, was observed in the ridge over the eastern United States prior to secondary cyclogenesis. This jet was located above the area of heavy snowfall associated with the inverted surface trough in the Tennessee Valley on 2 March. Wind speeds in the subtropical jet at the crest of the ridge did not increase significantly with time.

- Wind speeds near the base of the trough associated with the sea-level cyclone increased by 20 m s$^{-1}$ during the 24-hour period prior to 1200 UTC 3 March, as secondary cyclogenesis commenced near the Carolina coast. The region downwind of the 500-mb trough axis was transformed from a jet streak entrance region to a diffluent exit region by 1200 UTC 3 March, as a wind maximum formed or propagated to a position near the base of the trough. Secondary cyclogenesis occurred along the East Coast in the diffluent exit region of the intense upper-level jet located over the southeastern United States on 3 March.

- Missing wind reports at 200 and 300 mb at 0000 UTC and 1200 UTC 4 March may have indicated very high winds upstream from the trough along the East Coast.

## 7.5     10–13 DECEMBER 1960

*General remarks*

- This early-season storm was the first of three big snowstorms during the 1960/1961 winter season. It was accompanied by wind gusts of up to 42 m s$^{-1}$ at Block Island, Rhode Island, and 38 m s$^{-1}$ at Nantucket, Massachusetts. Temperatures falling below $-7°C$ created blizzard conditions across parts of the Middle Atlantic and New England states.
- Regions with snow accumulations exceeding 25 cm: West Virginia panhandle, extreme northern Virginia, Maryland (except for the lower eastern shore), parts of Delaware, southern and eastern Pennsylvania, New Jersey, southeastern New York, Connecticut, Rhode Island, Massachusetts (except the extreme west), extreme southeastern Vermont, southern New Hampshire, and coastal Maine.
- Regions with snow accumulations exceeding 50 cm: scattered locations in northern New Jersey and eastern Massachusetts.
- Urban center snowfall amounts:

  | | |
  |---|---|
  | Washington, D.C.–National Airport | 8.5″ (22 cm) |
  | Baltimore, MD | 14.1″ (36 cm) |
  | Philadelphia, PA | 14.6″ (37 cm) |
  | New York, NY–The Battery | 17.0″ (44 cm) |
  | Boston, MA | 13.0″ (33 cm) |

- Other selected snowfall amounts.

  | | |
  |---|---|
  | Newark, NJ | 20.4″ (52 cm) |
  | Trenton, NJ | 16.6″ (42 cm) |
  | Portland, ME | 14.9″ (38 cm) |
  | Hartford, CT | 13.4″ (34 cm) |
  | Providence, RI | 11.2″ (28 cm) |

*The surface chart*

- An Arctic front passed through the northeastern states on 10–11 December and was followed by colder weather as rising sea-level pressures extended eastward into New England from an anticyclone north of Minnesota (1035 mb).
- The inverted sea-level pressure ridge along the East Coast on 11 December indicates that cold air damming was occurring immediately prior to the formation of a coastal front and secondary cyclogenesis late on 11 December.
- The primary low developed in the western Gulf of Mexico by 1200 UTC 10 December. It propagated northward to Oklahoma, then eastward across Kentucky to West Virginia by 0000 UTC 12 December. The low center deepened slowly from 1012 to 1002 mb during this period. This system was associated with light to moderate precipitation amounts, with snow accumulations of generally less than 12 cm across the Plains states, Midwest, and Ohio Valley states. The primary low was observed for only 6 hours after the onset of secondary cyclogenesis at 1800 UTC 11 December.

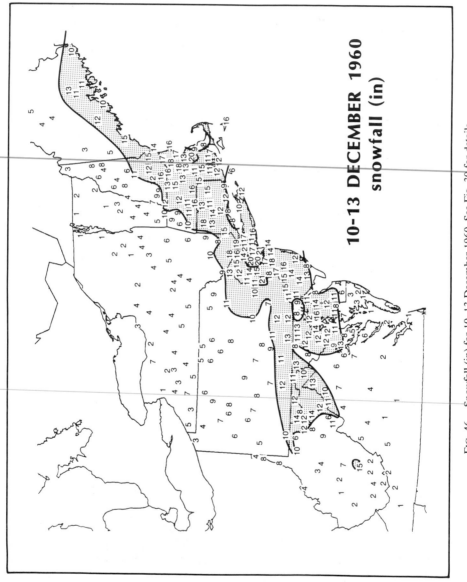

FIG. 46.   Snowfall (in) for 10–13 December 1960. See Fig. 30 for details.

# SURFACE

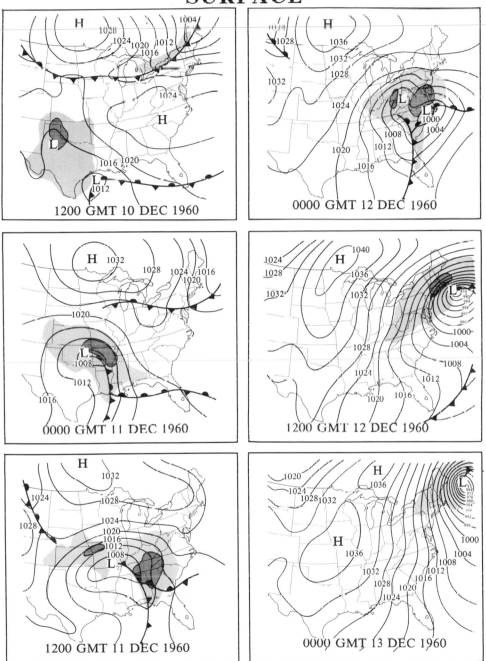

FIG. 47. Twelve-hourly surface weather analyses for 1200 UTC 10 December 1960 through 0000 UTC 13 December 1960. See Fig. 31 for details.

- The secondary low formed over South Carolina along a rapidly evolving coastal front. The low moved off the Atlantic coast after passing near Cape Hatteras, North Carolina, at 0000 UTC 12 December, advancing northeastward at approximately 16 to 17 m s$^{-1}$. Heavy snow fell from Virginia to New England as the secondary cyclone propagated from eastern North Carolina to a position south of Nantucket, Massachusetts, between 0000 and 1200 UTC 12 December. The central sea-level pressure dropped 27 mb during this 12-hour period and reached a minimum of 966 mb by 0000 UTC 13 December.
- Sea-level pressure gradients north of the low center intensified rapidly as the cyclone deepened on 12 December. Strong north-northeasterly winds were reported along the coastline.
- Extreme cold followed the storm, with surface temperatures across the eastern United States remaining 10°C below seasonal levels for the following few days.

## The 850-mb chart

- Strong lower-tropospheric cold advection was observed across the northeastern United States through 1200 UTC 11 December in association with rising sea-level pressures beneath a region of pronounced upper-level confluence.
- The 850-mb low center intensified and its circulation expanded late on 10 December, as the primary surface low deepened over Oklahoma. The 850-mb low strengthened slowly during 11 December as it tracked from Oklahoma to West Virginia. It then either redeveloped or moved rapidly toward the coast and deepened explosively by 1200 UTC 12 December, about 18 hours after secondary sea-level cyclogenesis was initiated.
- Prior to 0000 UTC 12 December, the largest low-level temperature gradients were concentrated in the cold air over the northeastern United States, while the thermal gradient across the southeastern United States remained fairly weak. The temperature gradient south of the 0°C isotherm increased during 11 December as warm advection, and a southerly 20–25 m s$^{-1}$ LLJ formed over the southeastern United States. The 0°C isotherm remained nearly stationary across West Virginia and Virginia prior to and during secondary cyclogenesis over North Carolina, despite significant warm air advection between 1200 UTC 11 December and 0000 UTC 12 December.
- Very cold air plunged southward from central Canada as two upper-level troughs merged during 12 December.
- An S-shaped isotherm pattern became established along the East Coast by 0000 UTC 12 December as secondary cyclogenesis commenced. The S-shaped signature developed as strong warming occurred along the Southeast coast, while temperatures remained relatively constant in the northeastern United States.
- Strong south to southwesterly low-level winds developed across the southeastern United States beneath the diffluent exit region of the upper-level jet system crossing the Gulf states on 11 December. The high wind speeds accelerated the moisture transport into the precipitation area which expanded over the eastern United States by 0000 UTC 12 December.

# 850 MB

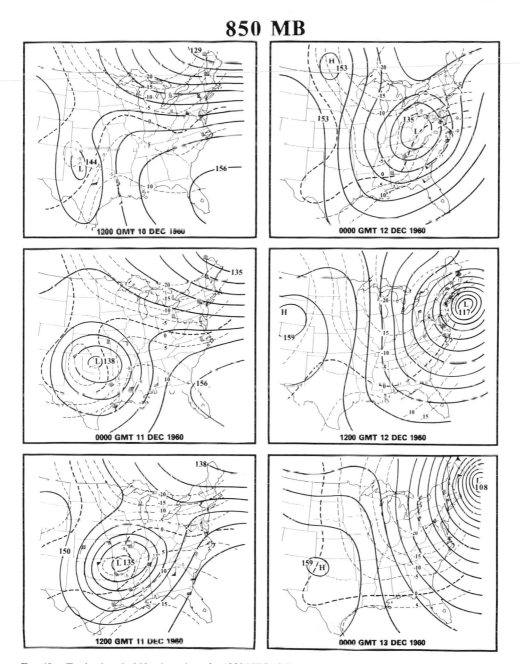

FIG. 48.    Twelve-hourly 850-mb analyses for 1200 UTC 10 December 1960 through 0000 UTC 13 December 1960. See Fig. 32 for details.

- Strong southeasterly winds were observed at 0000 UTC 12 December in advance of the 850-mb low center, over portions of the Middle Atlantic states where heavy snowfall was occurring. Large easterly wind components were found north of the 850-mb center from 0000 UTC 11 December through 1200 UTC 12 December, and were associated with moderate to heavy precipitation at each map time.

*500-mb geopotential height analyses*

- An intense vortex was evident over eastern Canada prior to and during the early stages of East Coast cyclogenesis. Cyclonic flow extended from central Canada across the northeastern United States on 10 and 11 December. The strong cyclonic flow combined with a ridge near the Southeast coast to produce pronounced confluence over the eastern United States between 1200 UTC 10 December and 1200 UTC 11 December. Very cold air and high pressure at the surface extended across the northeastern United States beneath this confluence zone.
- A high-amplitude ridge was located over the western United States and Canada, upstream of the closed 500-mb trough in the southwestern United States that was associated with the primary surface low-pressure center.
- The evolution of the sea-level cyclone appears to have been influenced primarily by the propagation of the southwestern trough to the east. As the cyclone neared the East Coast, however, it also interacted with a separate trough that propagated southeastward from Canada. Although this trough was difficult to find initially in the strong cyclonic flow over Canada, it was plainly visible from western Quebec to Wisconsin at 0000 UTC 12 December.
- The southwestern 500-mb trough was characterized by closed contours through 1200 UTC 11 December, but "opened up" as it neared the East Coast and merged with the other trough propagating southward from central Canada. This apparent interaction occurred as the secondary cyclone deepened rapidly along the East Coast.
- The two troughs were marked by distinct cyclonic vorticity maxima. Secondary cyclogenesis and the development of heavy snowfall along the East Coast were heavily influenced by the cyclonic vorticity advections associated with these two features; but the vorticity maximum associated with the trough that originated over the southwestern United States appears to have exerted the most important direct influence.
- The merged troughs formed a deep closed-contour low center over the northeastern United States by 0000 UTC 13 December.
- Amplitude changes were difficult to assess due to the transformations from vortex to open trough to vortex, but the Canadian trough that propagated southward and the downstream ridge amplified sharply after 0000 UTC 12 December, as rapid cyclogenesis continued off the Atlantic coast.
- The half-wavelength between the 500-mb center and the downstream ridge decreased prior to primary cyclogenesis between 1200 UTC 10 December and 0000 UTC 11 December. It also appeared to decrease during the rapid secondary cyclogenesis on 12 December.

# 500 MB HEIGHTS AND UPPER-LEVEL WINDS

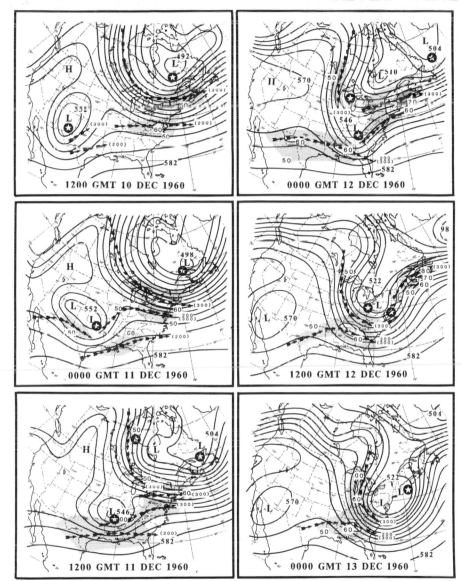

FIG. 49.   Twelve-hourly analyses of 500-mb geopotential heights and upper-level wind fields for 1200 UTC 10 December 1960 through 0000 UTC 13 December 1960. See Fig. 33 for details.

*Upper-level wind analyses*

- Intense polar jet streaks were located over the northeastern United States in association with highly confluent flow, as the cold sea-level high-pressure ridge extended across this region prior to cyclogenesis. Wind speeds in the polar jet over New England and eastern Canada increased by 15–20 m s$^{-1}$ between 1200 UTC 11 December and 1200 UTC 12 December, as the secondary cyclone formed and began to deepen rapidly along the East Coast.

- South of the initially closed trough over the southwestern United States, wind speeds of greater than 50 m s$^{-1}$ expanded in areal coverage, with maximum winds in this jet increasing to greater than 60 m s$^{-1}$ by 1200 UTC 11 December. The increases in wind speed and areal coverage occurred as the primary low and heavy precipitation developed in the southern United States.
- Another distinct jet streak was associated with the trough propagating southeastward across central Canada, with 50 m s$^{-1}$ speeds at 1200 UTC 11 December and 0000 UTC 12 December.
- The jets associated with these two separate troughs appeared to merge over the southeastern United States at 1200 UTC 12 December. The rapidly developing coastal storm was located beneath the diffluent exit region of this jet system off the East Coast.

## 7.6   18–20 JANUARY 1961

*General remarks*

- The "Kennedy Inaugural Snowstorm" occurred on the eve of John F. Kennedy's Presidential inauguration in Washington, D.C., and was the second of three major East Coast winter storms during the 1960/1961 season. The area of heaviest snowfall was similar in location to that in the December 1960 storm. Blizzard or near-blizzard conditions developed across the northeastern United States as the cyclone deepened rapidly offshore.
- Regions with snow accumulations exceeding 25 cm: northern Maryland and Delaware, parts of West Virginia and Virginia, southeastern Pennsylvania, New Jersey, southeastern New York, Connecticut, Rhode Island, Massachusetts (except for the northwest corner), and southern sections of New Hampshire and Maine.
- Regions with snow accumulations exceeding 50 cm: scattered parts of eastern Pennsylvania, northern New Jersey, southeastern New York, northwestern Connecticut, northeastern Massachusetts, and southern New Hampshire.
- Urban center snowfall amounts:

  | | |
  |---|---|
  | Washington, D.C.–National Airport | 7.7" (20 cm) |
  | Baltimore, MD | 8.4" (21 cm) |
  | Philadelphia, PA | 13.2" (34 cm) |
  | New York, NY–Central Park | 9.9" (25 cm) |
  | Boston, MA | 12.3" (31 cm) |

- Other selected snowfall amounts:

  | | |
  |---|---|
  | Harrisburg, PA | 18.7" (47 cm) |
  | Worcester, MA | 18.7" (47 cm) |
  | Nantucket, MA | 16.0" (41 cm) |

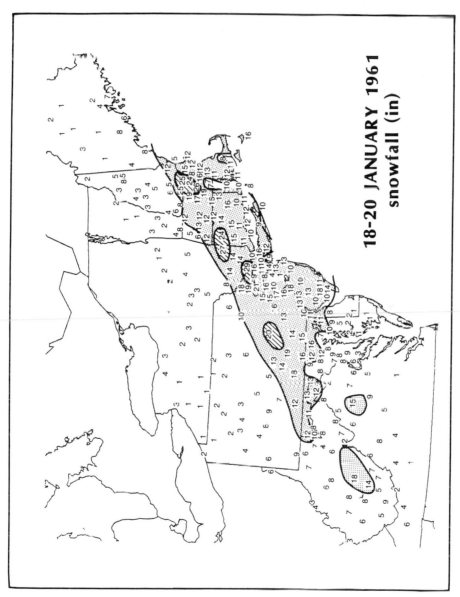

**18-20 JANUARY 1961**
**snowfall (in)**

FIG. 50.   Snowfall (in) for 18–20 January 1961. See Fig. 30 for details.

| | |
|---|---|
| Hartford, CT | 14.2″ (36 cm) |
| Newark, NJ | 13.7″ (35 cm) |

*The surface chart*

- A front ushered in cold air for the heavy snow event across New England and the Middle Atlantic states on 18 January. This cold air was associated with high pressure north of the Great Lakes.
- There was no evidence of cold air damming from New England into the Middle Atlantic states immediately prior to cyclogenesis on 19 January. No inverted sea-level pressure ridge was observed, and surface temperatures did not show the characteristic wedge of cold air along the eastern slopes of the Appalachians.
- The surface low appeared to follow a relatively straight path, even across the Appalachian range, as it moved toward the Atlantic coast on 19 January. Although no separate secondary cyclone was observed, the surface low did have a tendency to "jump" or reform slightly further to the east as it crossed the Appalachian Mountains.
- The cyclone propagated rapidly, covering approximately 700 to 800 km every 12 hours. The forward movement of the storm may have slowed slightly near the New England coast.
- There was no evidence of coastal frontogenesis, which could be related, in part, to both the lack of cold air damming and the speed with which this system developed and moved across the East Coast.
- The cyclone deepened rapidly from 1200 UTC 19 January through 0000 UTC 21 January. During this time, the central pressure fell 43 mb over one 24-hour period. The lowest sea-level pressure was analyzed at 964 mb east of New England late on 20 January. The intensification late on 19 January and early on 20 January was accompanied by a large increase in the pressure gradients surrounding the cyclone center and an increase in wind speeds along the coast.
- The rapid intensification of the surface low center coincided with an expansion of the precipitation shield on 19 January. Virtually no precipitation was occurring at 0000 UTC 19 January, with light snow breaking out across the Ohio Valley by 1200 UTC 19 January. Precipitation rates increased dramatically over the following 12 to 24 hours from the Middle Atlantic states northeastward into New England.
- Cold air was reinforced following the storm as a cold anticyclone plunged south-southeastward into the Plains states late on 20 January.

*The 850-mb chart*

- Northwesterly flow and cold advection were established over the northeastern United States between 1200 UTC 18 January and 1200 UTC 19 January, beneath a region of upper-level confluence. The 0°C isotherm remained virtually stationary near the Virginia–North Carolina border through 19 January.
- The 850-mb low commenced deepening between 0000 UTC and 1200 UTC 19 January [$-90$ m $(12$ h$)^{-1}$] over the Ohio Valley. It intensified most rapidly between 0000 and 1200 UTC 20 January [$-210$ m $(12$ h$)^{-1}$], at the same time that the surface low deepened by 24 mb along the East Coast.

# SURFACE

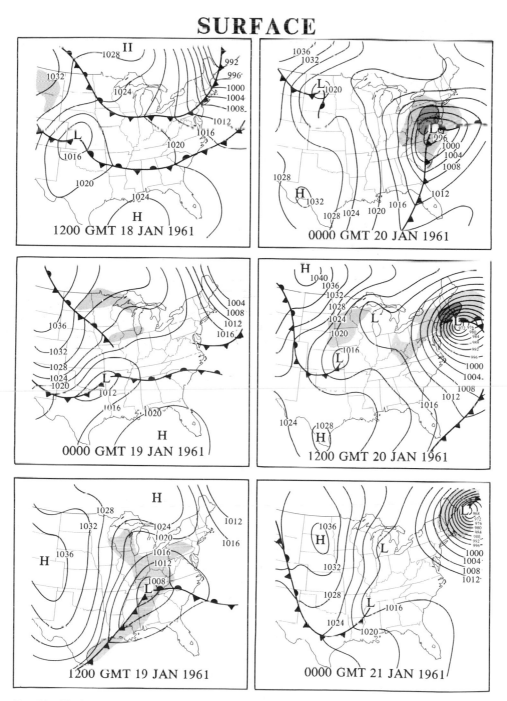

FIG. 51. Twelve-hourly surface weather analyses for 1200 UTC 18 January 1961 through 0000 UTC 21 January 1961. See Fig. 31 for details.

- The initial development of the 850-mb low over the central Plains states on 18 January was characterized by strong cold air advection west of the center. A northerly 25 m s$^{-1}$ LLJ was located beneath the entrance region of an upper-level jet across the southern Plains states (see upper-level wind analyses). East of the 850-mb low, only weak southwesterly flow and slight warm air advection were evident prior to 1200 UTC 19 January, when little precipitation was reported.

- Increasing warm air advection east of the deepening 850-mb low at 1200 UTC 19 January and 0000 UTC 20 January was associated with the widespread outbreak of precipitation. The strong cold air advection behind the low combined with the developing warm air advection ahead of it to produce an S-shaped isotherm pattern along the East Coast by 0000 UTC 20 January. This coincided with the start of the cyclone's explosive deepening stage. Temperature gradients were concentrated around the 850-mb low at this time, with the 0°C isotherm nearly collocated with the low center.

- An easterly wind component developed to the north and northeast of the 850-mb center after 1200 UTC 19 January as the 850-mb and surface lows began to deepen. By 1200 UTC 20 January, wind speeds had increased by 20–25 m s$^{-1}$ north of the low center, probably contributing to the intensification of snowfall rates during this phase of the storm's development.

*500-mb geopotential height analyses*

- The precyclogenetic period featured a high-amplitude ridge anchored over the west coast of North America. The trough that was associated with the cyclone was located downstream of this ridge. It was imbedded in a belt of enhanced geopotential height gradients extending from central Canada across the northern United States north of a ridge over the Gulf of Mexico.

- A slow-moving trough in eastern Canada was associated with a confluent geopotential height pattern across the Great Lakes and New England during 19 January. This upper-level configuration was located above the surface anticyclone north of the Great Lakes that was supplying cold air to the northeastern United States.

- The 500-mb trough associated with the cyclone propagated eastward across the United States on 19 and 20 January and evolved into a deep closed center by 0000 UTC 21 January, as the surface low intensified rapidly off the New England coast. The rapid development of the cyclone and precipitation shield occurred to the east of an amplifying cyclonic vorticity center, in the region of strong vorticity advection. This feature propagated from the Rocky Mountain states to Oklahoma, then up the Tennessee Valley, and finally off the Middle Atlantic coast on 18–20 January. The geopotential height gradient at the base of the trough increased at 0000 UTC 20 January, concurrent with the onset of sea-level development.

- As this system approached the East Coast, diffluence increased and the trough axis assumed a northwest–southeast orientation (a negative tilt) by 1200 UTC 20 January. The sea-level cyclone was deepening rapidly off southeastern New England at this time.

# 850 MB

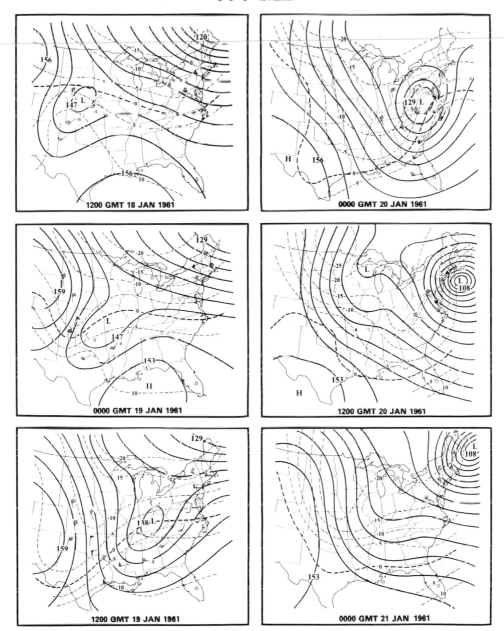

FIG. 52.   Twelve-hourly 850-mb analyses for 1200 UTC 18 January 1961 through 0000 UTC 21 January 1961. See Fig. 32 for details.

# 500-MB HEIGHTS AND UPPER-LEVEL WINDS

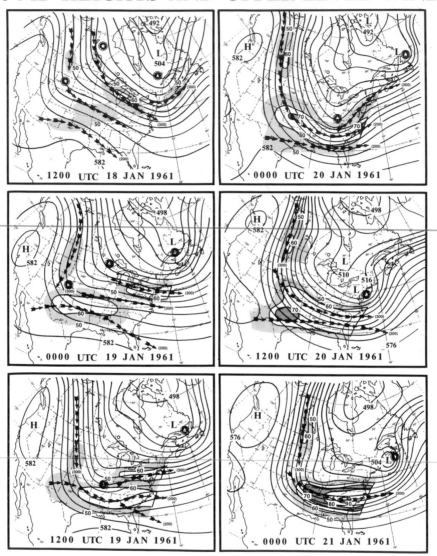

FIG. 53. Twelve-hourly analyses of 500-mb geopotential heights and upper-level wind fields for 1200 UTC 18 January 1961 through 0000 UTC 21 January 1961. See Fig. 33 for details.

- The amplitude of the trough over the central United States and the ridge *upstream* along the West Coast increased dramatically in the 24-hour period ending at 1200 UTC 19 January, as the surface low moved from Kansas to Tennessee. The amplitude increased slowly thereafter. The increase of amplitude was associated with the development of an upper-level anticyclone along the West Coast.

- As the surface low began to deepen on its trek from Tennessee to the Atlantic Ocean between 1200 UTC 19 January and 1200 UTC 20 January, the 500-mb geopotential heights east of the cyclone rose significantly. The rising heights produced a well-defined ridge by 1200 UTC 20 January and caused an increase in the amplitude of the trough and *downstream* ridge during the period when the surface low intensified rapidly.

- The half-wavelength between the trough and its *downstream* ridge over the eastern United States decreased after 0000 UTC 19 January, especially during the period of rapid cyclogenesis on 20 January.

*Upper-level wind analyses*

- Prior to cyclogenesis, confluent flow and a 60 m s$^{-1}$ polar jet streak extended from the Great Lakes across New Jersey and offshore between 1200 UTC 18 January and 1200 UTC 19 January. A cold surface high-pressure ridge built from the Great Lakes region into New England during this period.

- At 1200 UTC 18 January, 50 m s$^{-1}$ wind speeds were found at 200 and 300 mb over Montana, upwind of the trough that was later associated with the cyclogenesis off the East Coast. A broader area downwind of the trough contained similar wind speeds. By 0000 UTC 19 January, winds had increased to greater than 60 m s$^{-1}$ from near the base of trough eastward, with evidence of multiple jet streak structure. This region of high wind speeds expanded eastward by 1200 UTC 19 January, near the developing ridge axis over the East Coast. At the same time, the surface low and expanding precipitation shield were located over the Ohio and Tennessee valleys.

- By 0000 UTC 20 January, 200-mb wind speeds had increased to 70 m s$^{-1}$ both upwind and downwind of the trough axis, as the half-wavelength between the trough and downstream ridge decreased and the geopotential height gradients increased. The rapid cyclogenesis over Virginia around 0000 UTC 20 January occurred in the diffluent exit region of the amplifying jet streak that extended offshore from the Carolina coast and in the entrance region of a separate jet embedded within the confluent flow over southern New England.

- Winds strengthened dramatically upstream of the trough axis by 0000 UTC 21 January, with speeds exceeding 90 m s$^{-1}$ over the southeastern United States as the surface low moved east of New England.

## 7.7    2–5 FEBRUARY 1961

*General remarks*

- The third major snowstorm of the 1960/1961 winter season occurred at the end of one of the more prolonged cold spells experienced across the northeastern United States. It produced near-record snow cover in the major metropolitan areas since snow fell on unmelted accumulations from the previous storms. This storm also produced paralyzing gale- to hurricane-force winds on the coast, with wind gusts of 43 m s$^{-1}$ at Blue Hill Observatory, Milton,

Massachusetts, 41 m s$^{-1}$ at Block Island, Rhode Island, and 37 m s$^{-1}$ at New York City (La Guardia Airport). Temperatures rose to near freezing during the storm, producing heavy, wet snow accumulations along the coast.

- Regions with snow accumulations exceeding 25 cm: panhandle of West Virginia, parts of northern Virginia and northern Maryland, Pennsylvania, central and southern New York, central and northern New Jersey, Connecticut, Rhode Island, Massachusetts, southern Vermont, and New Hampshire.
- Regions with snow accumulations exceeding 50 cm: northern Pennsylvania, central New York, extreme southeastern New York, Long Island, northern New Jersey, scattered portions of New England.
- Urban center snowfall amounts:

| | |
|---|---|
| Washington, D.C.–National Airport | 8.3″ (21 cm) |
| Baltimore, MD | 10.7″ (27 cm) |
| Philadelphia, PA | 10.3″ (26 cm) |
| New York, NY–Kennedy Airport | 24.0″ (61 cm) |
| New York, NY–La Guardia Airport | 19.0″ (48 cm) |
| Boston, MA | 14.4″ (37 cm) |

- Other selected snowfall amounts:

| | |
|---|---|
| Cortland, NY | 40.0″ (102 cm) |
| Newark, NJ | 22.6″ (57 cm) |
| Worcester, MA | 18.8″ (46 cm) |
| Providence, RI | 18.3″ (45 cm) |
| Allentown, PA | 17.3″ (44 cm) |
| Nantucket, MA | 14.4″ (37 cm) |
| New Haven, CT | 14.0″ (36 cm) |

*The surface chart*

- A large, very cold anticyclone (1048 mb) north of the Great Lakes preceded the storm, with record low temperatures in the northeastern United States on 2 February. The anticyclone remained north of New York and New England prior to and during the snowstorm, serving as a continuous source of cold air for the snowfall in the Northeast.
- Cold air damming occurred on 2–3 February, with a distinct inverted ridge of high pressure extending from New York to northern Florida at 0000 and 1200 UTC 3 February.
- A coastal front developed off the southeast United States on 3 February, immediately prior to the development of the secondary cyclone.
- The primary low-pressure center intensified only slowly as it propagated through the Tennessee Valley on 2 February, then filled as it reached the Ohio Valley on 3 February. The central pressure of this system never fell below 1010 mb. The primary low remained a separate entity for 9 hours following the onset of secondary cyclogenesis along the East Coast on 3 February.
- The secondary low formed east of South Carolina shortly after 1200 UTC 3 February and deepened rapidly off the Middle Atlantic coast on 4 February. Intensification occurred primarily between 1200 UTC 3 February and 1200 UTC 4 February. The central sea-level pressure fell from 1008 mb near Nor-

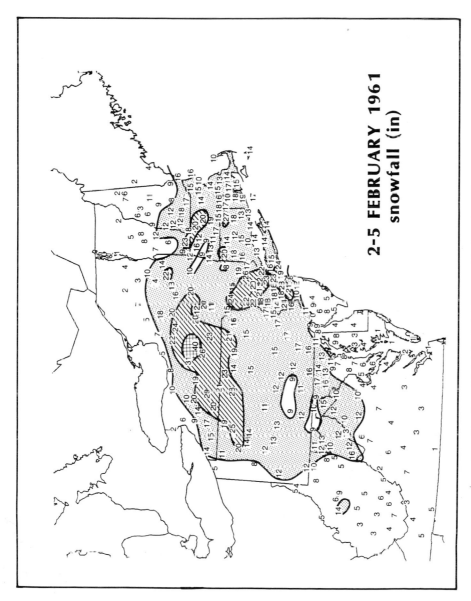

FIG. 54.  Snowfall (in) for 2–5 February 1961. See Fig. 30 for details.

# SURFACE

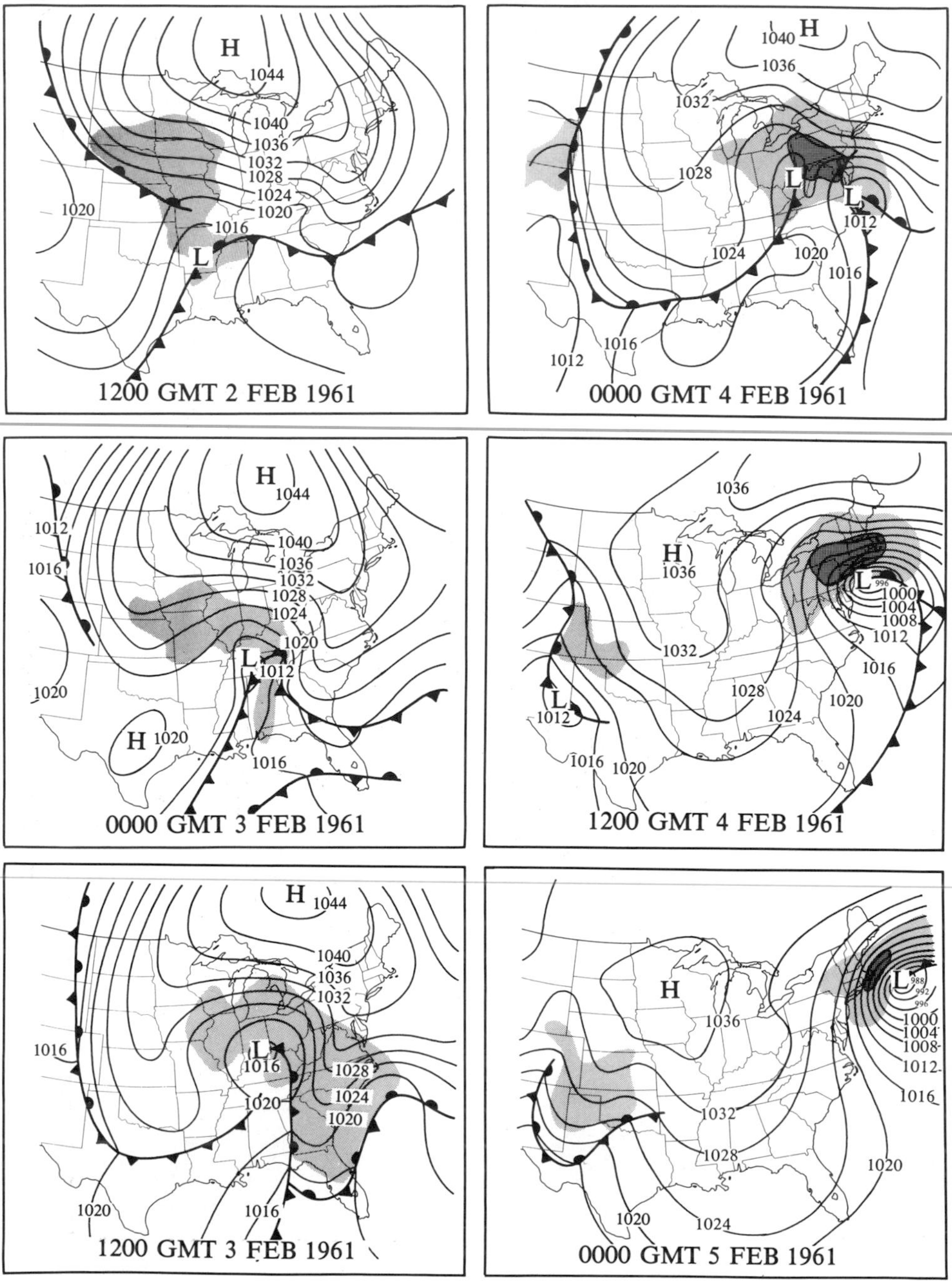

FIG. 55. Twelve-hourly surface weather analyses for 1200 UTC 2 February 1961 through 0000 UTC 5 February 1961. See Fig. 31 for details.

folk, Virginia, at 0000 UTC 4 February to 992 mb off the New Jersey coast 12 hours later. The secondary center moved northeastward parallel to the coast through 1200 UTC 4 February, then turned to the east during the next 12-hour period.

- Large sea-level pressure gradients formed to the north of the rapidly deepening secondary low on 4 February as heavy snow and strong northeasterly winds developed from northern Virginia to southern New England.
- The primary and secondary lows were both relatively slow movers, averaging only 12 to 13 m s$^{-1}$.

*The 850-mb chart*

- An 850-mb anticyclone located north of the Great Lakes was responsible for northwesterly flow and cold advection over the Northeast and Middle Atlantic states on 2 February. The cold air advection was located beneath a large region of upper-level confluence.
- The 850-mb low center located over the central United States on 2 February deepened slowly until 0000 UTC 3 February. This system propagated eastward and weakened slightly by 0000 UTC 4 February, then deepened explosively in the 12 hours ending at 1200 UTC 4 February [−240 m (12 h)$^{-1}$] as the surface cyclone was intensifying rapidly off the Middle Atlantic coast. Dual 850-mb low centers at 0000 UTC 4 February reflected the coastal redevelopment of the surface low center. Local height falls amplified markedly by 1200 UTC 4 February concurrent with the rapid deepening of the 850-mb low center.
- The 850-mb low evolved along a low-level baroclinic zone that extended from the central Plains states to the Middle Atlantic coast on 2–3 February. At 1200 UTC 2 February, the largest temperature gradients were oriented from west to east on the cold side of the 0°C isotherm. A pattern of cold advection west of the low and warm advection east of the low became quite evident by 0000 UTC 3 February. This established the S-shaped isotherm pattern along the East Coast during 3 February, as the secondary cyclone was forming. The strong warm air advection was close to the low center at 0000 UTC 3 February, but shifted to the coast during 3 February as coastal frontogenesis and cyclogenesis occurred along the Southeast coast.
- A southeasterly LLJ with wind speeds exceeding 25 m s$^{-1}$ formed over the Middle Atlantic states by 0000 UTC 4 February, in conjunction with secondary cyclogenesis, as the local height falls shifted from the Midwest to the East Coast. This jet appeared to enhance the moisture transport into the developing area of heavy precipitation along the East Coast.

*500-mb geopotential height analyses*

- The precyclogenetic environment over the northeastern United States was dominated by a large cyclonic circulation that drifted eastward across southeastern Canada. This vortex was associated with a region of pronounced confluence over eastern Canada and the northeastern United States, directly above the cold surface high that extended southeastward from the Great Lakes region.

# 850 MB

FIG. 56.   Twelve-hourly 850-mb analyses for 1200 UTC 2 February 1961 through 0000 UTC 5 February 1961. See Fig. 32 for details.

# 500 MB HEIGHTS AND UPPER-LEVEL WINDS

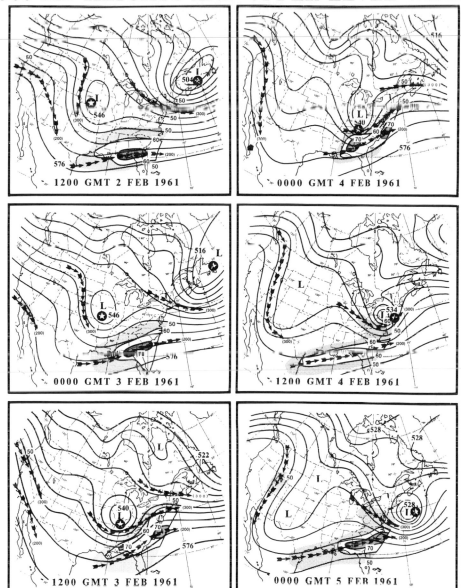

FIG. 57. Twelve-hourly analyses of 500-mb geopotential heights and upper-level wind fields for 1200 UTC 2 February 1961 through 0000 UTC 5 February 1961. See Fig. 33 for details.

- The sea-level cyclone was associated with a closed 500-mb low that drifted slowly eastward from the central Plains states to the East Coast from 2 to 4 February. This circulation was collocated with a well-defined cyclonic vorticity maximum and was associated with strong vorticity advections.
- The closed cyclonic circulation was evident over the Plains states prior to any major development at the surface. It deepened by 120 m between 0000 UTC 4 February and 0000 UTC 5 February, as the secondary cyclone developed rapidly off the East Coast. Geopotential height gradients increased at the base of the trough during this period.
- Intense sea-level development occurred between 0000 and 1200 UTC 4 February, as diffluence downwind of the trough crossed the coastline and the trough axis acquired a northwest–southeast orientation (a negative tilt).
- This storm was associated with an upper-level trough and downstream ridge of relatively small amplitude and short half-wavelength, as compared with other cases. A small increase in amplitude between the trough and downstream ridge occurred as the sea-level cyclone deepened after 0000 UTC 4 February. The half-wavelength between trough and downstream ridge decreased between 1200 UTC 3 February and 1200 UTC 4 February, as the trough approached the East Coast.

*Upper-level wind analyses*

- The entrance of a 50 m s$^{-1}$ polar jet within the confluent region over the northeastern United States coincided with the cold surface ridge which extended from New England to the Middle Atlantic states on 2 and 3 February.
- A subtropical jet with maximum wind speeds exceeding 70 m s$^{-1}$ at 200 mb was located in the downstream ridge along the Southeast coast through 3 February.
- A polar jet with maximum wind speeds of less than 50 m s$^{-1}$ propagated into the base of the trough by 0000 UTC 3 February as the primary low developed just east of the trough axis and precipitation expanded into the Ohio Valley. The possible merger of the polar and subtropical jets at 1200 UTC 3 February appeared to shift the axis of maximum wind speeds toward the base of the trough, concurrent with an increase in the geopotential height gradients there. The possible interaction of subtropical and polar jets at and after 1200 UTC 3 February yielded a complex jet pattern at 200 and 300 mb. Two distinct bands of strong wind speeds were observed downwind of the trough axis at separate levels and locations. The secondary cyclone developed between these two jets at 0000 UTC 4 February.
- The development of the secondary cyclone after 0000 UTC 4 February occurred in the diffluent exit region of the jet streaks located near the base of the trough.

## 7.8   11–14 JANUARY 1964

*General remarks*

- This system was a large, slow-moving storm that produced severe winter weather through much of the central and eastern United States. Blizzard conditions prevailed throughout the Middle Atlantic states and southern New England, as temperatures fell below −7°C and wind speeds increased to greater than gale force. The heaviest snows fell across Pennsylvania, where amounts of 40 to 60 cm were common.

- Regions with snow accumulations exceeding 25 cm: central and northern West Virginia, northern Maryland, the northern tip of Virginia, Pennsylvania (except the extreme northwest and southeast), New Jersey, the southeastern half of New York, portions of Connecticut and Massachusetts, Rhode Island, southern Vermont, southern New Hampshire, and the southern tip of Maine.
- Regions with snow accumulations exceeding 50 cm: portions of Pennsylvania and south-central New York.
- Urban center snowfall amounts:

| | |
|---|---|
| Washington, D.C. Dulles Airport | 10.2″ (26 cm) |
| Washington, D.C.–National Airport | 8.5″ (22 cm) |
| Baltimore, MD | 9.9″ (25 cm) |
| Philadelphia, PA | 7.2″ (18 cm) |
| New York, NY–Central Park | 12.5″ (32 cm) |
| Boston, MA | 9.2″ (23 cm) |

- Other selected snowfall amounts:

| | |
|---|---|
| Williamsport, PA | 24.1″ (61 cm) |
| Scranton, PA | 21.1″ (54 cm) |
| Nantucket, MA | 19.2″ (49 cm) |
| Harrisburg, PA | 18.1″ (46 cm) |
| Pittsburgh, PA | 15.6″ (40 cm) |
| Hempstead, NY | 14.7″ (37 cm) |

## The surface chart

- A large, intense anticyclone was located near the Minnesota–Canada border at 0000 UTC 12 January with a central pressure of 1046 mb. Ridges extended south from the northern Plains to Texas and southeast across the Great Lakes, then southward along the Atlantic coast. Cold temperatures were established across the northeastern states on 11–12 January as sea-level pressures rose over the Middle Atlantic states and New England. This high-pressure ridge remained north of New York during the storm period, serving as a continuous source of cold air along the coast.
- Cold air damming was in evidence along much of the East Coast at 1200 UTC 12 January and 0000 UTC 13 January. The characteristic inverted ridge of high pressure extended from the Middle Atlantic states into the southeastern states.
- A slowly intensifying primary low pressure center tracked from Kansas across Tennessee to Kentucky before weakening over western Virginia by 1200 UTC 13 January, 24 hours after the onset of secondary cyclogenesis along the East Coast. The primary low was accompanied by significant snowfall from Mis-

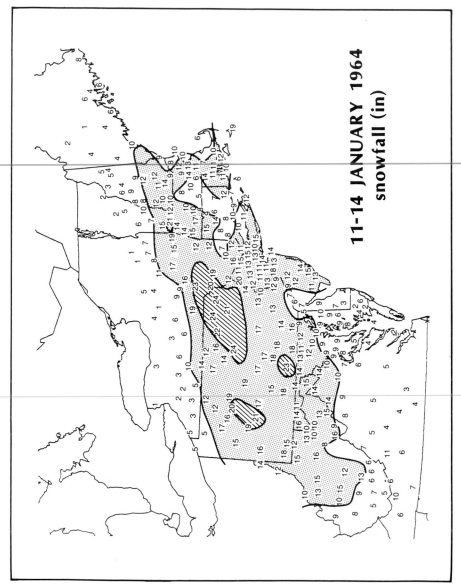

FIG. 58.    Snowfall (in) for 11–14 January 1964. See Fig. 30 for details.

# SURFACE

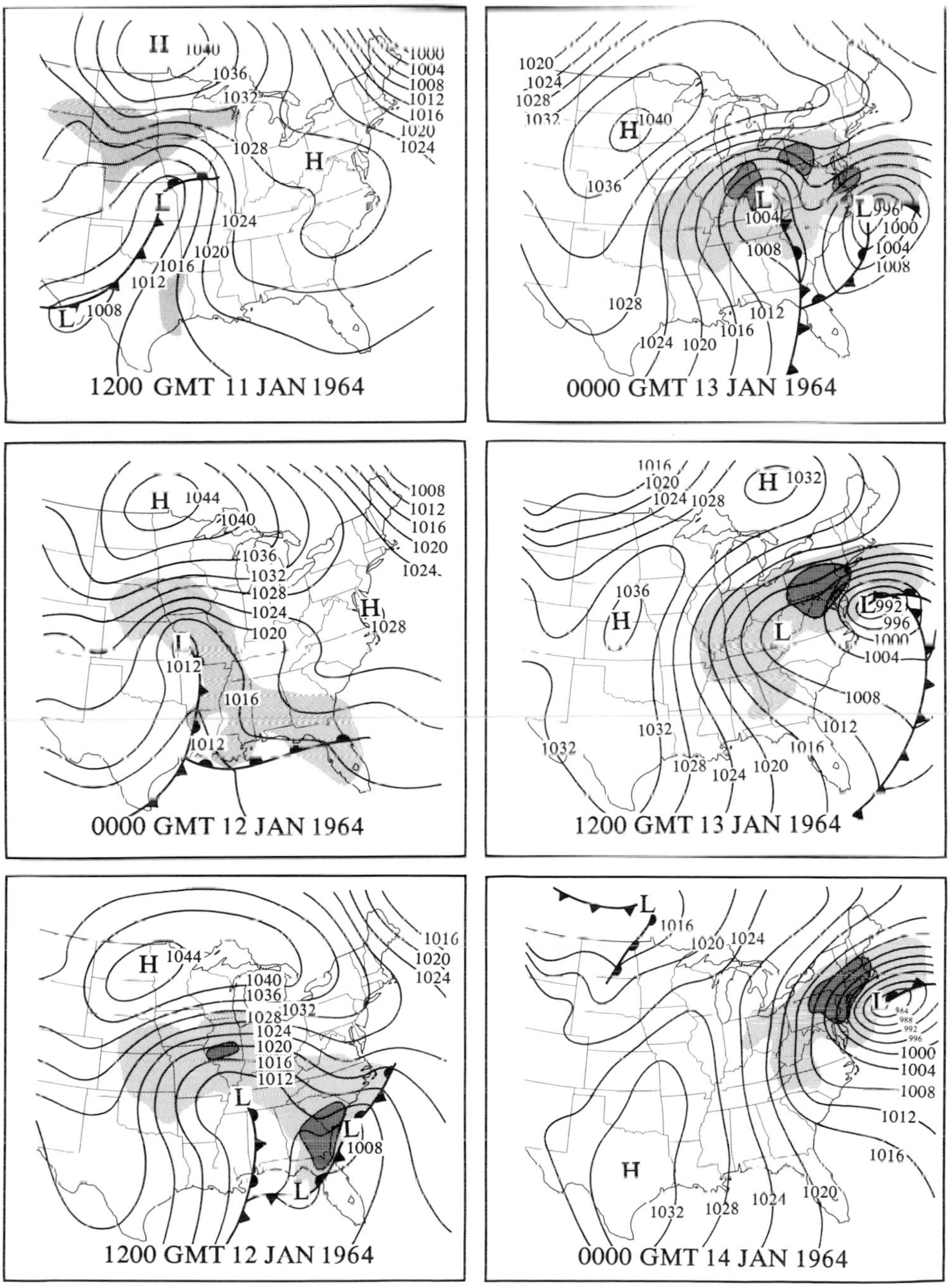

FIG. 59.   Twelve-hourly surface weather analyses for 1200 UTC 11 January 1964 through 0000 UTC 14 January 1964. See Fig. 31 for details.

souri and Iowa up the Ohio Valley into Pennsylvania, with accumulations exceeding 25 cm.

- A secondary cyclone developed as a series of low centers over the southeastern United States consolidated into one cyclone center off the North Carolina coast late on 12 January. The secondary cyclogenesis commenced before 1200 UTC 12 January during a period which featured coastal frontogenesis, cold air damming, and the development of moderate to heavy precipitation west of the coastal front.
- The secondary cyclone had an erratic history of intensification. The low deepened rapidly between 1200 UTC 12 January and 0000 UTC 13 January, did not deepen at all early on 13 January, then deepened rapidly again later on 13 January off the southern New England coast.
- Both the primary and secondary low-pressure centers moved at a relatively slow rate, averaging 9–12 m s$^{-1}$. The slow movement and erratic intensification created many problems for forecasters concerned with the onset time and duration of the snowfall.

### The 850-mb chart

- Northwesterly flow and cold air advection were located beneath a region of upper-level confluence over the northeastern United States between 1200 UTC 11 January and 1200 UTC 12 January. The cold northwesterly flow across New England persisted until 1200 UTC 13 January.
- The 850-mb low center drifted eastward from Kansas to West Virginia and deepened very slowly from 11 through 13 January. A secondary 850-mb low began to form near the South Carolina–Georgia border at 1200 UTC 12 January. This center deepened at a more rapid pace than the primary low, as it propagated northeastward along the East Coast. Between 1200 UTC 13 January and 0000 UTC 14 January, the two 850-mb low centers consolidated into one strong center near the coast.
- The development of a secondary 850-mb low center during 12 January was associated with the formation of a 20 m s$^{-1}$ south-to-southeasterly LLJ along the Florida to South Carolina coast. These developments coincided with the secondary sea-level cyclogenesis and coastal frontogenesis off the Southeast coast, and the outbreak of moderate to heavy precipitation across Georgia and South Carolina.
- The 0°C isotherm was displaced to the south and east of the initial 850-mb low on 11–12 January and was later located to the north of the secondary 850-mb center at 0000 UTC 13 January. The east-to-west orientation of the 850-mb isotherms at 1200 UTC 11 January evolved into an S-shaped pattern by 0000 UTC 12 January as cold air was advected southward behind the 850-mb low and warm air was drawn northward ahead of it.
- The temperature gradients increased along the Middle Atlantic coast during 12 January with the onset of secondary cyclogenesis, coastal frontogenesis, and the formation of the LLJ.
- A strong easterly jet was established to the north of both the primary and

# 850 MB

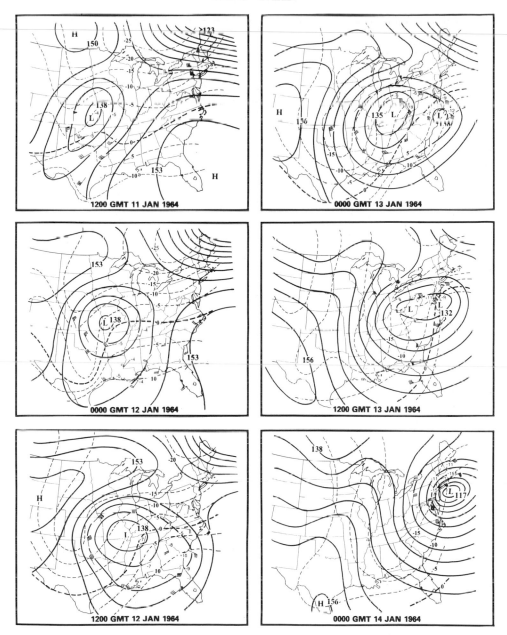

FIG. 60.   Twelve-hourly 850-mb analyses for 1200 UTC 11 January 1964 through 0000 UTC 14 January 1964. See Fig. 32 for details.

secondary 850-mb low centers on 12, 13, and 14 January. This LLJ strengthened as the 850-mb low deepened rapidly by 0000 UTC 14 January, with wind speeds exceeding 25 m s$^{-1}$.

- The development of a coastal height-fall center and LLJ over the 12-hour period ending at 1200 UTC 12 January was similar to that which occurred in the February 1979 "Presidents' Day Storm" (Bosart 1981; Uccellini et al. 1984, 1987). The LLJ and height-fall center formed at a considerable distance from the primary 850-mb low located over the central United States at 1200 UTC 12 January. These features developed at the 850-mb level while a subtropical jet strengthened at 200 mb, cold air damming occurred east of the Appalachian Mountains, and coastal frontogenesis and heavy precipitation were developing near the Southeast coast. The LLJ increased the thermal and moisture advections across the Middle Atlantic states. The January 1964 cyclone was of a much larger scale than the 1979 storm, and the LLJ developed with a better-defined coastal height-fall center than that found in the 1979 case.

*500-mb geopotential height analyses*

- An intense anticyclonic circulation over Greenland and a strong cyclonic vortex over eastern Canada established a region of confluent geopotential heights across the northeastern United States from 11 January through 1200 UTC 13 January. This confluent flow regime was located above the rising sea-level pressures that accompanied a surge of cold air from the Great Lakes to the Middle Atlantic coast on 12 January.
- A separate cyclonic vortex was located in the central United States on 12 January, downstream from a high-amplitude ridge over the western United States. The rather intense 500-mb circulation did not appear to result from the occlusion of an earlier cyclone and persisted throughout the period of this storm. The large-amplitude trough associated with this vortex drifted across the central United States as the primary surface low developed in Kansas and moved slowly to the Ohio Valley on 11 and 12 January. The circulation center deepened from 5300 m at 1200 UTC 12 January to 5120 m by 0000 UTC 14 January, as the secondary surface low formed along the East Coast.
- An intense area of cyclonic vorticity was collocated with the vortex in association with cyclonic wind shear at the base of the trough. The sea-level cyclone developed off the East Coast in a region of cyclonic vorticity advection east of the vortex.
- Diffluence was observed downwind of the trough from 12 through 14 January.
- The amplitude of the trough and its flanking ridges did not appear to change significantly immediately prior to or during secondary cyclogenesis. The amplitude of the trough and downstream ridge increased only slightly throughout the study period.
- The half-wavelength between the trough and downstream ridge decreased only slightly, primarily between 1200 UTC 11 January and 0000 UTC 13 January, when secondary surface cyclogenesis was under way.

# 500-MB HEIGHTS AND UPPER-LEVEL WINDS

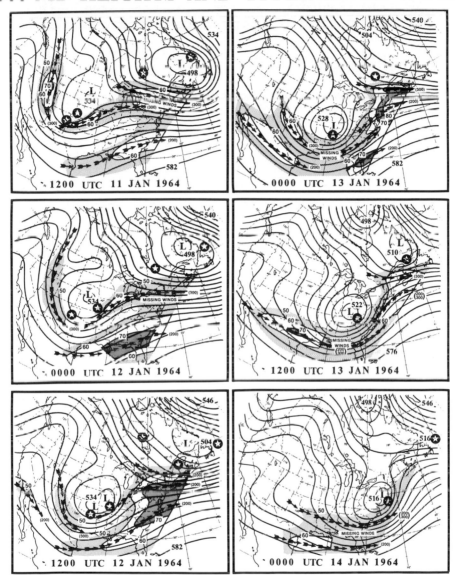

FIG. 61. Twelve-hourly analyses of 500-mb geopotential heights and upper-level wind fields for 1200 UTC 11 January 1964 through 0000 UTC 14 January 1964. See Fig. 33 for details.

*Upper-level wind analyses*

- Missing wind reports over the northeastern United States at 1200 UTC 11 January and 0000 UTC 12 January were indicative of the strong winds associated with a highly confluent polar jet. During this period, high sea-level pressure and cold air began to surge into the northeastern United States.

- An intensifying subtropical jet, with maximum winds exceeding 70 m s$^{-1}$ at 200 mb, was located over the southeastern United States between 1200 UTC 11 January and 0000 UTC 12 January, prior to the development of the secondary surface low off the Southeast coast.
- Wind speeds at 200 mb increased significantly over the Middle Atlantic states and New England in the 12-hour period ending at 1200 UTC 12 January. This jet strengthened to greater than 80 m s$^{-1}$ over New England by 0000 UTC 13 January. This feature developed and amplified at the crest of the downstream ridge as a coastal front, secondary cyclone, and heavy precipitation were developing over the southeastern United States. The amplification of both upper-level and low-level jet streaks along the Southeast coast (see the 850-mb maps), in conjunction 'with the development of heavy precipitation in Georgia and South Carolina at 1200 UTC 12 January, is similar to the sequence observed in the February 1979 "Presidents' Day Storm" (Uccellini et al. 1984, 1987).
- A polar jet with maximum wind speeds of at least 60 m s$^{-1}$ propagated into the base of the trough over the east-central United States by 0000 UTC 13 January, as the development of the secondary cyclone was initiated near the East Coast. Missing wind reports in the base of the trough at 0000 UTC and 1200 UTC 13 January indicate that very high winds probably developed within this jet streak as it propagated toward the Middle Atlantic coast while secondary cyclogenesis was in progress just offshore.
- The secondary low developed in the diffluent exit region of the complex jet system just east of the trough axis and in the entrance region of the confluent jet over New England.

## 7.9   29–31 JANUARY 1966

*General remarks*

- This storm is referred to as the "Blizzard of '66." The storm was the third and most severe in a series of three snowstorms that occurred over a 10-day period along the Middle Atlantic coast. Heavy snow fell from North Carolina northward to New York and combined with temperatures below −10°C and wind gusts in excess of 25 m s$^{-1}$ to create widespread blizzard conditions. Cyclone snowfall combined with "lake effect" (see section 3.1) snow over parts of New York state to produce record accumulations of 150 to 250 cm immediately south and east of Lake Ontario. The axis of heaviest snowfall extended from southwest to northeast across Virginia, but shifted to a north–south orientation north of Virginia. It was displaced west of the immediate coast from New Jersey to New England, sparing the major metropolitan areas from Philadelphia across much of southern New England the storm's worst effects.
- Regions with snow accumulations exceeding 25 cm: parts of North Carolina, central Virginia, parts of West Virginia, Maryland (except the extreme west), Delaware, northwestern New Jersey, central and eastern Pennsylvania, New

York (except the northeast, southeast, and parts of the southwest), and scattered locations throughout New England.

- Regions with snow accumulations exceeding 50 cm. central New York.
- Urban center snowfall amounts:

| | |
|---|---|
| Washington, D.C.–National Airport | 13.8" (35 cm) |
| Baltimore, MD | 12.1" (31 cm) |
| Philadelphia, PA | 8.3" (21 cm) |
| New York, NY–Central Park | 6.8" (17 cm) |
| Boston, MA | 6.3" (16 cm) |

- Selected snowfall amounts:

| | |
|---|---|
| Syracuse, NY | 39.0" (99 cm)[21] |
| Rochester, NY | 26.7" (68 cm) |
| Binghamton, NY | 20.1" (51 cm) |
| Roanoke, VA | 12.3" (31 cm) |
| Harrisburg, PA | 12.2" (31 cm) |

*The surface chart*

- The cyclone developed along the advancing edge of a record-setting cold air mass associated with a 1055-mb anticyclone over northern Canada.
- Cold air damming was observed early on 29 January in conjunction with an inverted high-pressure ridge near the Southeast coast. Concurrently, a coastal front developed just offshore.
- The surface low-pressure center initially moved along the Gulf coast at a rate of about 15 m s$^{-1}$ before turning northeastward along the Southeast coast on 29 January. The low deepened slowly as it propagated northeastward along the coastal front through the afternoon of 29 January. The center tended to "jump" or redevelop northeastward along the coastal front near the Carolina coast late on 29 January. It advanced at 25 m s$^{-1}$ as it crossed eastern North Carolina by 0000 UTC 30 January.
- Once the cyclone center passed eastern North Carolina, it took a more northerly track late on 29 January, deepening explosively from 0000 UTC 30 January until 1800 UTC 30 January. The central pressure fell from 997 to 970 mb. Very heavy snow fell from Virginia to New York during this stage of rapid development. The subsequent track of the storm took it inland across extreme eastern New Jersey and New York City, then up the Hudson Valley of New York. To the west of this path, temperatures never rose much above −10°C, and north-northwest surface winds strengthened, creating blizzard conditions from central Virginia to central New York. A wedge of drier and warmer air raced up the coast near and to the east of the low center. Temperatures approached 5°C in New York City and southern New England, cutting snow accumulations in these regions, since snowfall ceased or changed to rain for a 6- to 9-hour period.
- Intense sea-level pressure gradients formed, especially to the west and north of the low center, prolonging high winds even after the low-pressure system had passed.

---

[21] Snowfall amounts for the N.Y. cities are accumulations through 31 January.

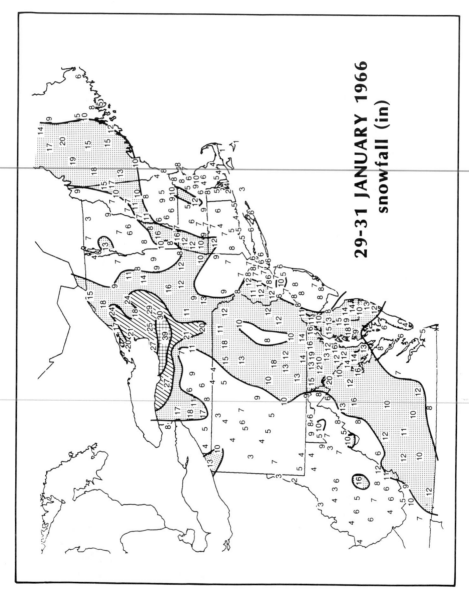

FIG. 62.   Snowfall (in) for 29–31 January 1966. See Fig. 30 for details.

# SURFACE

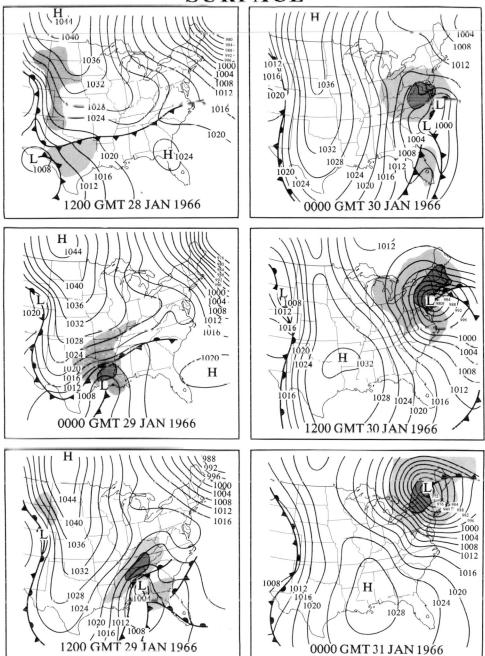

FIG. 63. Twelve-hourly surface weather analyses for 1200 UTC 28 January 1966 through 0000 UTC 31 January 1966. See Fig. 31 for details.

*The 850-mb chart*

- A strong northwesterly flow of very cold air was maintained across much of the Northeast through 1200 UTC 29 January, suppressing the 0°C isotherm southward to the Gulf states. Most of the northern states were covered by 850-mb temperatures below −20°C.
- The 850-mb low intensified only slightly through 0000 UTC 29 January, then deepened at a rate of −60 m $(12\ h)^{-1}$ over the 24-hour period ending at 0000 UTC 30 January, as the surface low reached North Carolina. During the following 12-hour period, the surface low intensified explosively and the 850-mb center also deepened rapidly [−150 m $(12\ h)^{-1}$]. A slower rate of intensification [−60 m $(12\ h)^{-1}$] was then observed in the 12-hour period ending at 0000 UTC 31 January.
- The 850-mb low developed along the southern edge of an intense baroclinic zone that extended from the southern Plains to the Middle Atlantic coast. The pronounced temperature gradient was maintained over the Middle Atlantic states within a confluent geopotential height pattern on 28 and 29 January. The east–west orientation of the thermal field along the East Coast on 28–29 January changed to north–south on 30 January. This change in orientation occurred during the rapid development phase of the storm and was marked by a 24°C temperature difference between Washington, D.C. (−24°C), and New York City (0°C) at 1200 UTC 30 January. The 0°C isotherm was located near the 850-mb low center, except during the occluding stage of the cyclone late on 30 January, when the center was found in the colder air over New England.
- An S-shaped isotherm pattern did not become clearly evident until 1200 UTC 30 January, during the period of explosive cyclogenesis. This pattern emerged as strong warm air advection developed along the Middle Atlantic coast by 0000 UTC 30 January, and across the northeastern states at 1200 UTC 30 January.
- On 28–29 January, during the early stages of cyclogenesis along the Gulf coast, a southerly 10–20 m s$^{-1}$ LLJ was directed toward areas of heavy precipitation in advance of the 850-mb low. Easterly flow was not observed to the north of the 850-mb low center until rapid cyclogenesis occurred on the Atlantic coast at 1200 UTC 30 January. A very strong southeasterly jet developed northeast of the 850-mb low after 0000 UTC 30 January, in association with rapid 850-mb deepening. Wind speeds of 30–40 m s$^{-1}$ were measured along the southern New England coast. This strong low-level flow was directed toward the region reporting heavy snowfall, indicating that the vertical motions and moisture transport associated with the LLJ were important in this case.

*500-mb geopotential height analyses*

- Prior to rapid cyclogenesis on 30 January, an amplifying ridge moved inland from the West Coast as a trough propagated out of the southwestern United States. An anticyclonic circulation between Quebec and Greenland moved

# 850 MB

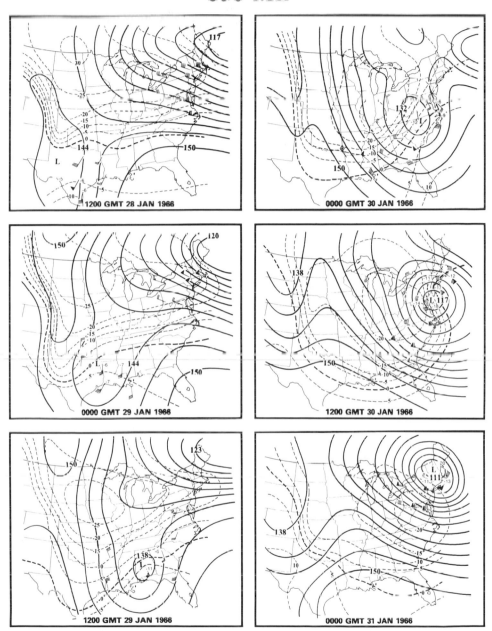

FIG. 64.    Twelve-hourly 850-mb analyses for 1200 UTC 28 January 1966 through 0000 UTC 31 January 1966. See Fig. 32 for details.

slowly westward, with an elongated cyclonic circulation extending from the Great Lakes eastward beyond Nova Scotia.

- This case was characterized by a major reorientation of the upper-level features. The propagating trough from the southwestern United States merged with a separate trough that propagated and amplified downwind of the ridge over the northwestern United States. This separate trough was marked by a distinct upper-level jet streak and an amplifying cyclonic vorticity maximum that tracked southeastward from Saskatchewan to Iowa between 1200 UTC 28 January and 1200 UTC 29 January. The merger of these features was associated with the splitting of the elongated closed low near the Canadian border into two separate vortices by 1200 UTC 29 January. One center moved eastward off the Newfoundland coast on 29 January. The other center, located near Lake Superior at 1200 UTC 29 January, rotated from an east–west to a north–south orientation by 0000 UTC 30 January.

- The surface low deepened explosively on 30 January as the vorticity maximum that had dropped southeastward from the northern Plains on 28–29 January neared the East Coast. The merger of the two troughs and the increasing cyclonic wind shear yielded one vorticity maximum near the North Carolina–Virginia coast by 1200 UTC 30 January.

- The rotation of the Great Lakes trough and its subsequent merger with the trough from the southwestern United States occurred as the amplitude of the combined trough system and upstream ridge increased substantially between 1200 UTC 28 January and 0000 UTC 30 January. The amplitude of the trough and the downstream ridge over the northeastern United States increased after 1200 UTC 29 January, as the surface low moved to the Southeast coast, then intensified rapidly.

- The maximum intensification of the surface low occurred as the diffluent region downwind of the trough became better defined and the trough axis became oriented from northwest to southeast (a negative tilt) by 1200 UTC 30 January, with the surface low approaching New York City. This region also contained the strongest vorticity advection downwind of the trough.

- The vortex deepened significantly between 1200 UTC 30 January and 0000 UTC 31 January. The half-wavelength of the trough and downstream ridge decreased between 0000 UTC 29 January and 0000 UTC 30 January as the surface low developed along the East Coast and heavy precipitation expanded in the southeastern and Middle Atlantic states.

*Upper-level wind analyses*

- A 60 m s$^{-1}$ polar jet streak was located over the northeastern United States on 28 January, immediately south of the elongated 500-mb vortex. This jet was associated with the outbreak of cold air across the eastern states prior to cyclogenesis. The extension of the colder air over the Middle Atlantic states by 1200 UTC 29 January occurred within the confluent region of the jet.

- Maximum wind speeds associated with the subtropical jet extending across the southeastern United States at 200 mb increased from 70 m s$^{-1}$ to greater than 80 m s$^{-1}$ between 1200 UTC 28 January and 1200 UTC 29 January,

# 500 MB HEIGHTS AND UPPER-LEVEL WINDS

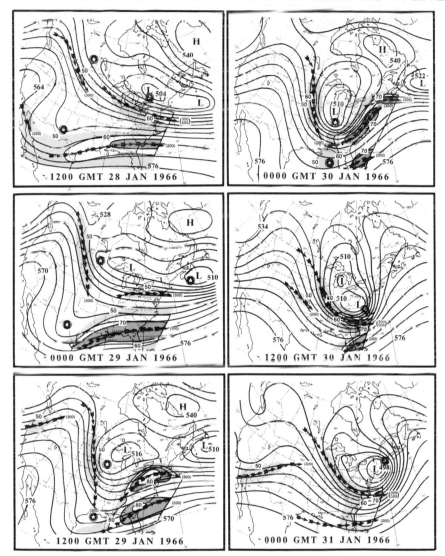

FIG. 65.    Twelve-hourly analyses of 500-mb geopotential heights and upper-level wind fields for 1200 UTC 28 January 1966 through 0000 UTC 31 January 1966. See Fig. 33 for details.

during the early stages of cyclogenesis. A coastal front developed along the Southeast coast and precipitation expanded across the Gulf coast region concurrent with this increase in wind speeds.

- The most remarkable wind feature was the development of an 80 m s$^{-1}$ jet streak at 300 mb across the Middle Atlantic and New England states by 1200 UTC 29 January. This jet developed at the crest of the downstream ridge prior to rapid cyclogenesis on the East Coast and before heavy snowfall enveloped the Middle Atlantic region by 0000 UTC 30 January.

- A polar jet streak with maximum wind speeds in excess of 50 m s$^{-1}$ propagated southeastward from the northern Plains to the Gulf coast between 1200 UTC 28 January and 1200 UTC 29 January as the West Coast ridge amplified and the Great Lakes vortex began to rotate into a north–south orientation.
- The strongest upper-level wind speeds were consistently located downstream of the major trough, although winds increased at the base of the trough as it amplified by 1200 UTC 30 January. The rapid cyclogenesis occurred in the diffluent exit region of the complex jet system that became better defined along the Southeast coast by 1200 UTC 30 January.

## 7.10   23–25 DECEMBER 1966

*General remarks*

- This case is notable as a Christmas Eve snowstorm which deposited heavy snow over a wide area extending from the southern Plains states to New England. An intriguing aspect of this storm was the numerous reports of thunderstorms with heavy snow from the Middle Atlantic states to New England. Snowfall along the coast was reduced by sleet during the afternoon of the 24th.
- Regions with snow accumulations exceeding 25 cm: portions of central and northern Virginia, parts of Maryland and West Virginia, northern Delaware, eastern Pennsylvania, western New Jersey, eastern New York, western New England.
- Regions with snow accumulations exceeding 50 cm: parts of eastern New York.
- Urban center snowfall amounts:

  | | |
  |---|---|
  | Washington, D.C.–Dulles Airport | 9.2″ (23 cm) |
  | Baltimore, MD | 8.5″ (22 cm) |
  | Philadelphia, PA | 12.7″ (32 cm) |
  | New York, NY–Central Park | 7.1″ (18 cm) |
  | Boston, MA | 5.7″ (14 cm) |

- Other selected snowfall amounts:

  | | |
  |---|---|
  | Albany, NY | 18.7″ (47 cm) |
  | Burlington, VT | 14.9″ (38 cm) |
  | Allentown, PA | 13.3″ (34 cm) |
  | Wilmington, DE | 12.5″ (32 cm) |
  | Trenton, NJ | 11.7″ (30 cm) |

*The surface chart*

- A strong high-pressure cell located over the Plains states was associated with a large mass of cold air covering the central and eastern United States on 22 and 23 December.

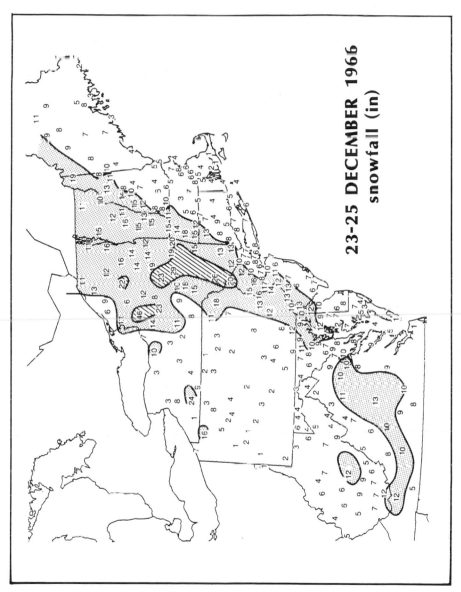

FIG. 66.   Snowfall (in) for 23–25 December 1966. See Fig. 30 for details.

- Cold air damming, as indicated by a weak inverted sea-level pressure ridge, was observed over the Middle Atlantic states between 0000 and 1200 UTC 24 December.
- Coastal frontogenesis was not observed along the Southeast coast since a preexisting frontal zone was already located there.
- The surface low developed along the edge of the cold air outbreak and moved eastward from Texas to the South Carolina coast by the morning of 24 December. The center deepened slowly during this period, while advancing at a pace of 15 m s$^{-1}$. The cyclone then moved northeastward just off the East Coast and deepened rapidly during the ensuing 24-hour period. The central pressure decreased 22 mb in 24 hours, reaching 978 mb by the time the low was located along the Maine coast at 1200 UTC 25 December.
- The surface low exhibited a "center jump" along the preexisting frontal boundary near the Carolina coast early on 24 December.

*The 850-mb chart*

- West to northwesterly flow over the northeastern United States maintained cold air in the lower troposphere into 24 December, with the 0°C isotherm located across southern Virginia.
- The 850-mb low center deepened only slightly as it moved across the southern United States on 23 December, but intensified more rapidly after 0000 UTC 24 December as it neared the East Coast. The 12-hourly central height changes were −60 m (12 h)$^{-1}$ at 1200 UTC 24 December, −90 m (12 h)$^{-1}$ at 0000 UTC 25 December, and −120 m (12 h)$^{-1}$ at 1200 UTC 25 December.
- Strong cold air advection west of the low center preceded the development of warm air advection east of the center. The cold air advection was observed by 0000 UTC 23 December, as strong winds developed nearly perpendicular to the band of isotherms stretched across Texas. A strong northerly jet to the rear of the 850-mb low dominated early cyclogenesis across the southern United States on 23 December. Otherwise, weak flow was observed near the low center prior to 24 December. Significant warm air advection was not seen until 0000 UTC 24 December as the 850-mb low propagated slowly eastward and southerly winds increased along the Southeast coast.
- The 850-mb low propagated along a narrow, but intense, east–west temperature gradient. The isotherms evolved into an S-shaped configuration between 0000 and 1200 UTC 24 December as the warm air advection pattern strengthened east of the low. The 0°C isotherm ran through the 850-mb low center until the cyclone occluded. By 1200 UTC 25 December, the low center was located in colder air over New England.
- Wind speeds increased near the low center by 1200 UTC 24 December as it neared the East Coast and began to intensify. A LLJ developed to the south and east of the low as the surface cyclone was beginning a 24-hour period of rapid deepening. Easterly flow strengthened north of the low by 0000 UTC 25 December, during rapid 850-mb intensification, and was directed toward the area of heavy snowfall.

# SURFACE

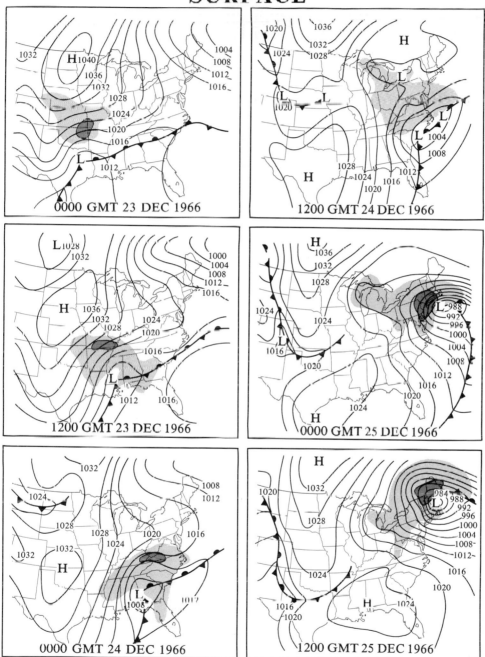

FIG. 67. Twelve-hourly surface weather analyses for 0000 UTC 23 December 1966 through 1200 UTC 25 December 1966. See Fig. 31 for details.

# 850 MB

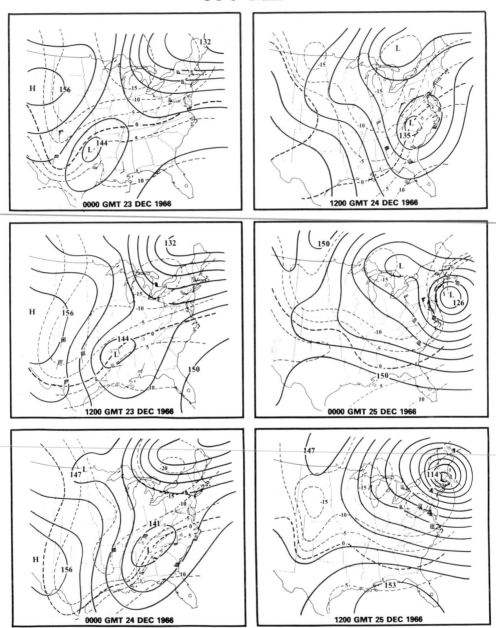

FIG. 68.     Twelve-hourly 850-mb analyses for 0000 UTC 23 December 1966 through 1200 UTC 25 December 1966. See Fig. 32 for details.

*500-mb geopotential height analyses*

- The precyclogenetic environment featured a ridge just inland from the West Coast and a trough extending from Nebraska to New Mexico early on 23 December. A closed cyclonic circulation was located just north of Lake Huron. An amplifying ridge over northeastern Canada was later associated with the development of a closed anticyclone which drifted slowly westward across northern Quebec.
- The juxtaposition of the Great Lakes vortex and the trough in the Plains states produced confluence across the northern and eastern United States. The confluent region appeared to influence the surface high-pressure ridge building eastward into the Middle Atlantic states from the anticyclone over the Plains states on 23 December.
- The development of the surface low occurred in conjunction with the eastward propagation of the trough initially over the central Plains states at 0000 UTC 23 December as it interacted with the cyclonic vortex over the Great Lakes by 1200 UTC 24 December.
- A cyclonic vorticity maximum propagated northeastward along the East Coast as the cyclone deepened on 24–25 December.
- The vortex over the Great Lakes on 23 December split into two separate trough systems early on 24 December. One trough moved eastward from Newfoundland and the other became established over the central Great Lakes, similar to the January 1966 case. As this occurred, the trough over the Plains states merged with the trough system forming near Michigan.
- A 500-mb low center remained over Michigan as a new center developed over Maryland at 0000 UTC 25 December, as the surface low deepened rapidly off the East Coast. This new center then pivoted counterclockwise around the Michigan low and was located over New England at 1200 UTC 25 December. During this period, it deepened by $-120$ m $(12$ h$)^{-1}$, while the surface low continued to intensify off the New England coast. The sea-level cyclone intensified beneath a region of increasing diffluence east of the trough, which attained a northwest–southeast orientation (a negative tilt) by 0000 UTC 25 December.
- The amplitude of the *upstream* ridge over the western United States and the trough over the central Plains increased sharply by 0000 UTC 24 December. Meanwhile, the trough drifted southeastward to the lower Mississippi Valley, prior to the intensification of the surface low along the East Coast. The amplitude of the trough and *downstream* ridge increased markedly over the 12-hour period ending at 0000 UTC 24 December, prior to East Coast cyclogenesis, and increased moderately thereafter as the cyclone deepened.
- The half-wavelength between the trough and downstream ridge decreased slightly between 0000 UTC 24 December and 0000 UTC 25 December during surface cyclogenesis and as the precipitation shield spread northeastward along the East Coast.

*Upper-level wind analyses*

- Wind speeds along the axes of the upper-level jets did not increase substantially in this case.

# 500-MB HEIGHTS AND UPPER-LEVEL WINDS

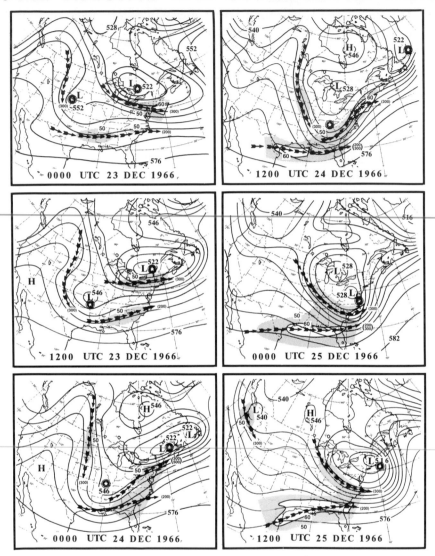

FIG. 69.   Twelve-hourly analyses of 500-mb geopotential heights and upper-level wind fields for 0000 UTC 23 December 1966 through 1200 UTC 25 December 1966. See Fig. 33 for details.

- A polar jet streak was observed over the northeastern United States on 23 December within the confluent region south of the vortex over the Great Lakes. The confluent entrance region of this jet was associated with rising sea-level pressures and colder temperatures extending into the Middle Atlantic states.
- A subtropical jet at 200 mb, located downwind of the trough over the central

United States, had maximum winds of 50 m s$^{-1}$. These speeds were less than those observed in many other cases.

- A new axis of strong winds developed from Mississippi to New Jersey, on the east side of the trough, at 0000 UTC 24 December. Wind speeds in this jet reached 60 m s$^{-1}$ over southern New England by 1200 UTC 24 December.
- Wind speeds exceeding 50 m s$^{-1}$ increased in areal coverage after 0000 UTC 24 December as precipitation expanded north and east of the developing low-pressure system over the southeastern United States. Wind speeds increased at the base of the trough and upstream after 0000 UTC 24 December, when two jet axes were identified. Between 0000 UTC 24 December and 0000 UTC 25 December, the surface low deepened beneath the diffluent exit region of the jet systems downwind of the trough.

## 7.11    5–7 FEBRUARY 1967

*General remarks*

- Snowfall was produced from two separate low-pressure systems. The first cyclone produced a narrow band of snow across the northern Middle Atlantic states and southern New England, with accumulations generally less than 10 cm. It also ushered very cold air into New England and the Middle Atlantic states. The second storm produced heavy snowfall rates, but for a relatively short duration. As this storm moved rapidly northeastward along the East Coast, blizzard conditions developed across the Middle Atlantic and southern New England states.
- Regions with snow accumulations exceeding 25 cm: northern Virginia, northeastern West Virginia, central and northern Maryland, northern Delaware, southeastern Pennsylvania, parts of western Pennsylvania, most of New Jersey, extreme southeastern New York, Connecticut, Rhode Island, central and eastern Massachusetts (except Cape Cod), southeastern New Hampshire, and coastal Maine.
- Regions with snow accumulations exceeding 50 cm: scattered locations in northern New Jersey and southeastern New York.
- Urban center snowfall amounts:

| | |
|---|---|
| Washington, D.C.–Dulles Airport | 11.8″ (30 cm) |
| Baltimore, MD | 10.6″ (27 cm) |
| Philadelphia, PA | 9.9″ (25 cm) |
| New York, NY–Central Park | 15.2″ (39 cm) |
| Boston, MA | 9.5″ (24 cm) |

- Other selected snowfall amounts:

| | |
|---|---|
| Newark, NJ | 16.5″ (42 cm) |
| Bridgeport, CT | 14.9″ (38 cm) |
| Worcester, MA | 14.4″ (37 cm) |
| Trenton, NJ | 13.8″ (35 cm) |
| Allentown, PA | 13.0″ (33 cm) |

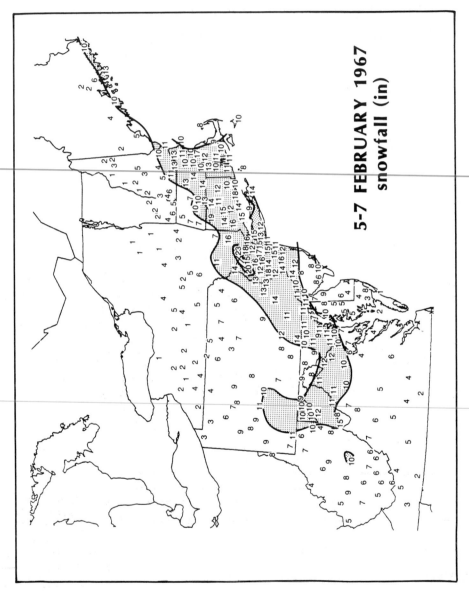

**5-7 FEBRUARY 1967 snowfall (in)**

Fɪɢ. 70.   Snowfall (in) for 5–7 February 1967. See Fig. 30 for details.

*The surface chart*

- Two cyclones were responsible for the heavy snowfall. The first sea-level low moved from southern Iowa across northern Kentucky to northern Virginia on 5–6 February. It propagated along a cold front that drifted southward into the Middle Atlantic states by 6 February. The surface low was weak and filled as it neared the East Coast, but still produced 5–10 cm accumulations from Pennsylvania into the New York City area.
- Following the first cyclone, an anticyclone (1040 mb) passed across southern Ontario into Quebec, accompanied by bitterly cold temperatures. A weak inverted sea-level pressure ridge west of the Middle Atlantic coast at 0000 and 1200 UTC 7 February is suggestive of cold air damming.
- Coastal frontogenesis was not observed in this case.
- The second low, which produced the heaviest snow, formed along the same cold front as it reached the Gulf Coast on 6 February. This cyclone moved very rapidly, propagating from northwestern Florida at 0000 UTC 7 February to a position off the Virginia coast 12 hours later. The low center appeared to "jump" from central South Carolina to the Virginia coast between 0900 and 1200 UTC 7 February. It then raced northeastward to a position off the coast of Maine by 0000 UTC 8 February, covering more than 1000 km in 12 hours.
- An area of precipitation expanded quickly to the east and north as the surface low moved rapidly up the Atlantic coast on 6–7 February. The heaviest precipitation was initially concentrated just north of the cyclone center, but was later observed approximately 200 km northwest of the low center. Heavy snow generally fell for only 12 hours or less due to the storm's rapid movement. Temperatures ranging between $-7°C$ and $-15°C$ combined with gale-force winds to create blizzard conditions across the Middle Atlantic and southern New England states on 7 February.
- The cyclone deepened slowly as it tracked across the southeastern United States. Deepening rates increased to nearly $-1$ mb h$^{-1}$ after 1200 UTC 7 February, as the storm center moved over the Atlantic Ocean. Although the deepening rates exhibited by this system were not as spectacular as those in many other cases, the storm's rapid advance resulted in local pressure falls across southeastern Canada that were as large as those observed in slower-moving cyclones that intensified more vigorously.

*The 850-mb chart*

- Separate 850-mb low centers accompanied the two surface low-pressure systems in this case. The first 850-mb low weakened as it moved up the Ohio Valley to Pennsylvania by 1200 UTC 6 February. Its subsequent passage off the East Coast was followed by the southward displacement of the 850-mb baroclinic zone to the Middle Atlantic region. This established cold air across the northeastern United States.
- The second and more significant 850-mb low center developed in Texas along the southern edge of a narrow, well-defined baroclinic zone that extended from Texas across the lower Ohio Valley to the Middle Atlantic states at 1200

# SURFACE

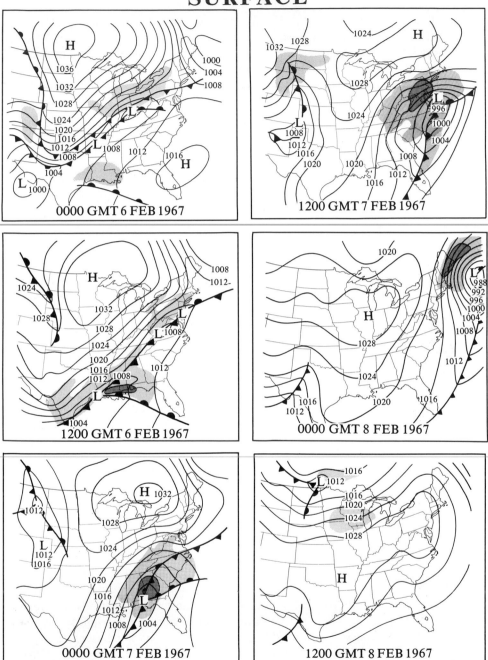

FIG. 71.   Twelve-hourly surface weather analyses for 0000 UTC 6 February 1967 through 1200 UTC 8 February 1967. See Fig. 31 for details.

# 850 MB

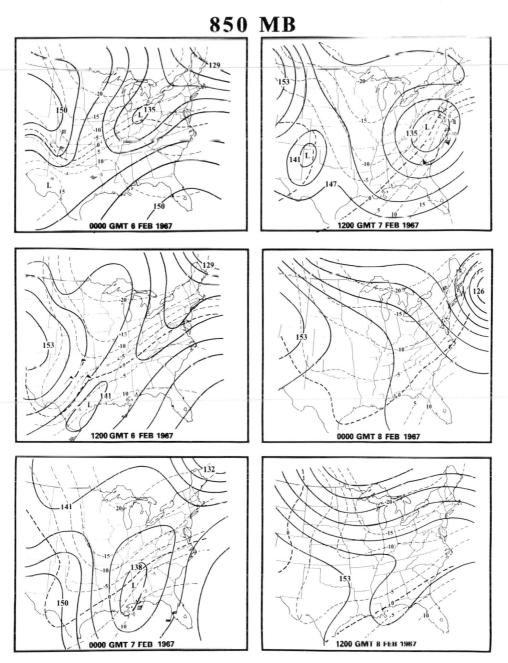

FIG. 72.　Twelve-hourly 850-mb analyses for 0000 UTC 6 February 1967 through 1200 UTC 8 February 1967. See Fig. 32 for details.

UTC 6 February. A 20–25 m s$^{-1}$ north-to-northeasterly jet accompanied the development of the 850-mb low over Texas on 6 February. Strong cold air advection occurred with this jet west of the newly formed center.

- The low deepened slowly as it propagated northeastward to the Middle Atlantic coast by 1200 UTC 7 February. It intensified more rapidly once it was well off the coast later on 7 February.

- The low center tracked along the narrow, intense band of isotherms which had been established prior to cyclogenesis. Only a hint of the S-shaped isotherm pattern appeared along the East Coast at 1200 UTC 7 February, a signature that was much better defined in many of the other cases.

- A 30 m s$^{-1}$ southerly jet formed over eastern North Carolina and wind speeds generally increased as the 850-mb low neared the Atlantic coast at 1200 UTC 7 February. The strengthening circulation enhanced the warm air advection and moisture transport over the Middle Atlantic states while moderate to heavy snowfall was developing in the region.

*500-mb geopotential height analyses*

- A deep trough propagating slowly across eastern Canada on 6 February produced highly confluent flow over southeastern Canada and the northeastern United States. An anticyclone with very cold surface temperatures moved toward New England on 6–7 February beneath the confluent upper-level flow.

- The first weakening cyclone that propagated to the south of New York City at 1200 UTC 6 February was associated with an ill-defined trough over the Ohio Valley. Although this trough was not easy to identify in the geopotential height field within the strongly confluent flow over the Ohio Valley and New England, there was an area of cyclonic vorticity advection over the Ohio Valley at 0000 UTC 6 February and over the Middle Atlantic states at 1200 UTC 6 February.

- A trough over northern Mexico and another trough propagating southeastward downwind of a high-amplitude ridge along the Pacific coast set the stage for the second, more intense cyclone on 6–7 February. As the trough became better defined over the central United States, a principal cyclonic vorticity maximum emerged over South Dakota at 1200 UTC 6 February, propagating to Illinois by 0000 UTC 7 February. The trough advanced rapidly to southeastern Canada by 1200 UTC 8 February, while the sea-level cyclone intensified along the East Coast.

- No closed cyclonic circulation developed at 500 mb and no distinct minimum in the geopotential height field was observed until 1200 UTC 8 February over extreme eastern Canada. Unlike most of the other cases, this system *did not* appear to have a marked diffluent region downstream of the trough axis until it reached eastern Canada, even though the trough developed a slight northwest–southeast (negative) tilt after 1200 UTC 7 February. The absence of this feature may be related to the rapid movement of the entire storm system. The geopotential height gradients increased at the base of the trough after 0000 UTC 7 February, during the period of rapid cyclogenesis.

- Large increases in amplitude were associated with geopotential height rises

# 500 MB HEIGHTS AND UPPER-LEVEL WINDS

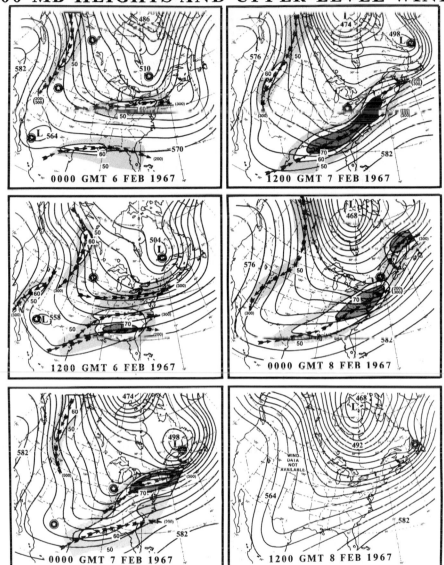

FIG. 73.   Twelve-hourly analyses of 500-mb geopotential heights and upper-level wind fields for 0000 UTC 6 February 1967 through 1200 UTC 8 February 1967. See Fig. 33 for details.

near the East Coast downstream of the trough at 0000 and 1200 UTC 7 February, as the cyclone was developing along the Southeast and Middle Atlantic coast. There was little change in the half-wavelength between the trough and downstream ridge until the cyclone was east of Newfoundland on 8 February.

*Upper-level wind analyses*

- A 60 m s$^{-1}$ polar jet streak was analyzed within the confluent region across the Ohio Valley and New England from 1200 UTC 5 February through 1200 UTC 6 February. This jet accompanied the first, weak surface low and was associated with the outbreak of cold air to the rear of this cyclone.
- Large increases in upper-level wind speeds were observed prior to and during cyclogenesis. Wind speeds within the subtropical jet over the southeastern United States increased by greater than 20 m s$^{-1}$ during the 24-hour period ending at 1200 UTC 6 February, as precipitation developed across the Gulf Coast. Wind speeds also strengthened over the northeastern United States by 0000 UTC 7 February as an 80 m s$^{-1}$ jet streak formed in the confluent flow at the crest of the upper-level ridge. This wind maximum developed as precipitation was expanding northward along the East Coast and the surface low was organizing along the Gulf Coast. Wind speeds at 200 and 300 mb increased uniformly to greater than 70 m s$^{-1}$ along the East Coast by 1200 UTC 7 February, as the cyclone moved off the Virginia coast and intensified rapidly.
- During 7 February, the surface low was located in the exit region of the jet to the south of the trough and the entrance region of the jet streak over New England and southeastern Canada, a pattern which is similar to that found in other cases.

## 7.12   8–10 FEBRUARY 1969

*General remarks*

- This cyclone is known as the "Lindsay Storm" in New York City since Mayor John Lindsay ran into political misfortune after sections of the city remained unplowed for a week following the snowfall. The storm was poorly predicted in New York City as forecasters first thought the precipitation would fall primarily as rain. Even when it appeared that snow would predominate on 9 February, predicted snowfall amounts persistently lagged the actual totals, as snow continued to fall long past the time it was forecast to cease. The rapid development and deceleration of the storm brought paralyzing snow and increasing winds from northern New Jersey through most of New England.
- Regions with snow accumulations exceeding 25 cm: northern New Jersey, southeastern New York (except eastern Long Island), most of Connecticut, central and northern Rhode Island, Massachusetts (except the southeast), Vermont (except the northwest), New Hampshire, and southern and central Maine.
- Regions with snow accumulations exceeding 50 cm: parts of the New York City and Boston metropolitan areas, western Connecticut, western and eastern Massachusetts, southern Vermont, northern Rhode Island, eastern New Hampshire, and southern Maine.
- Urban center snowfall amounts:

| | |
|---|---|
| Washington, D.C.–Dulles Airport | 5.0″ (13 cm) |
| Baltimore, MD | 3.0″ (8 cm) |

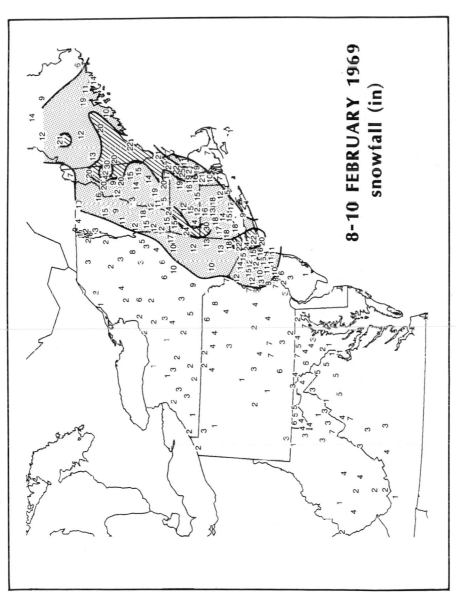

FIG. 74.   Snowfall (in) for 8–10 February 1959. See Fig. 50 for details

| | |
|---|---|
| Philadelphia, PA | 2.9″ (7 cm) |
| New York, NY–Kennedy Airport | 20.2″ (51 cm) |
| Boston, MA | 11.1″ (28 cm) |

- Other selected snowfall amounts:

| | |
|---|---|
| Bedford, MA | 25.0″ (64 cm) |
| Portland, ME | 21.5″ (55 cm) |
| Bridgeport, CT | 17.7″ (45 cm) |
| Hartford, CT | 15.8″ (40 cm) |
| Worcester, MA | 15.6″ (40 cm) |

*The surface chart*

- Bitterly cold air neither preceded nor followed the passage of this storm. This case was unusual in that no large, cold surface anticyclone was present over Ontario, Quebec, or northern New England immediately prior to East Coast cyclogenesis. Instead, a rather meager 1022 mb anticyclone drifted off New England by 0000 UTC 9 February, resulting in southeasterly winds along the coast late on 8 February. The winds backed to northeast along the Atlantic coast as the secondary low developed rapidly and moved east-northeastward off the Virginia coast on 9 February.

- Very weak cold air damming was observed along the Middle Atlantic coast on 8 February as the moderately cold high-pressure system crossed New England.

- The primary surface low propagated east-northeastward from Oklahoma to Kentucky in the 24-hour period ending at 0000 UTC 9 February, covering approximately 1000 km. Little change in the central sea-level pressure occurred during this period, but moderate to heavy rain was observed north of the low from Missouri to Ohio on 8 February. This low weakened rapidly as a secondary center developed along the East Coast between 0000 and 1200 UTC 9 February. The primary cyclone disappeared within 9 hours of the commencement of secondary cyclogenesis.

- The secondary low-pressure system formed over Georgia after 1800 UTC 8 February as heavy rains developed in the Carolinas. Mixed snow and rain spread across the Middle Atlantic states between 0000 and 0600 UTC 9 February, with wet snow predominating in northern Virginia, central Maryland, eastern Pennsylvania, and New Jersey. Generally light to moderate precipitation amounts were observed in these locations. Heavy snow developed from New Jersey northward by 1200 UTC 9 February.

- The secondary low developed inland over Georgia along the warm front of the primary low-pressure system. Although there was some tendency for this front to extend itself to the Southeast coast where small land–sea air temperature differences were observed, it cannot strictly be called a coastal front.

- The low-pressure center moved northeastward from the North Carolina coast to near Long Island by 0000 UTC 10 February, then eastward off Cape Cod by 1200 UTC 10 February. The low deepened 32 mb over an 18-hour period, reaching 970 mb by 0000 UTC 10 February just east of Long Island.

# SURFACE

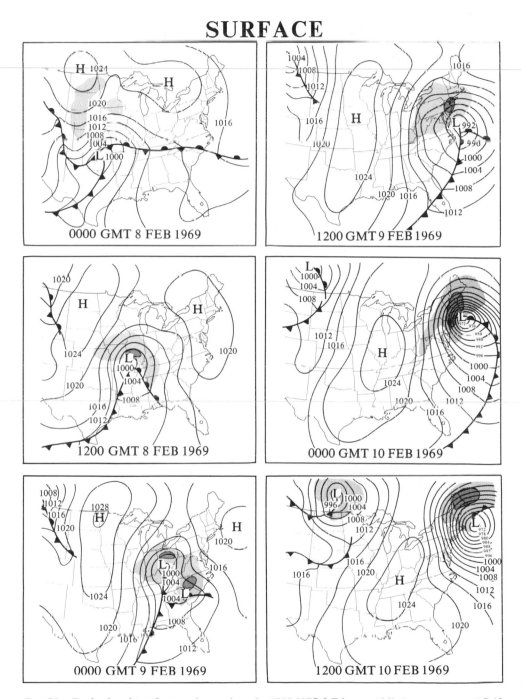

FIG. 75.  Twelve-hourly surface weather analyses for 0000 UTC 8 February 1969 through 1200 UTC 10 February 1969. See Fig. 31 for details.

- The forward motion of the secondary low slowed dramatically on 9 February, with the center advancing only 200 km in 12 hours. The storm's slow movement prolonged the snowfall from New York City across New England.

*The 850-mb chart*

- Very weak cold air advection was observed across the northeastern United States prior to East Coast cyclogenesis on 8 February. Lower-tropospheric temperatures ranged between 0°C and −10°C from Virginia to Maine.
- The 850-mb low was located at the edge of an intense temperature gradient over the southern Plains states at 0000 UTC 8 February, which weakened as the low drifted eastward in the following 12 hours. A tongue of warm air was located northeast and east of the 850-mb center on 8 February, allowing precipitation to fall as rain in the Ohio Valley.
- The 850-mb low began deepening after 1200 UTC 8 February as the primary surface low was crossing the Ohio Valley. The 850-mb deepening rate reached a maximum of −150 m (12 h)$^{-1}$ between 1200 UTC 9 February and 0000 UTC 10 February, concurrent with a 20-mb intensification of the secondary surface low off the Middle Atlantic coast.
- An elongated region of height falls from Ohio to South Carolina at 0000 UTC 9 February hinted at secondary 850-mb low development over the southeastern United States. Secondary development toward North Carolina was also supported by the large temperature gradient between Georgia and Virginia, well south and east of the main 850-mb low center.
- A pattern of strong cold air advection behind the 850-mb low emerged over Texas and Oklahoma by 1200 UTC 8 February, 12 hours prior to the onset of secondary cyclogenesis along the East Coast. Southwesterly winds increased over Louisiana, Mississippi, and Alabama, initiating the stronger warm air advection that later attended the early stages of secondary cyclogenesis along the Atlantic coast at 0000 UTC 9 February. The warm advection was greatest in the region with enhanced geopotential height falls and where precipitation was breaking out across the Carolinas.
- The isotherms began to exhibit an S-shape by 1200 UTC 9 February, at the start of rapid cyclogenesis. The 0°C isotherm was consistently located near the 850-mb low center.
- Easterly winds in the vicinity of the 850-mb low strengthened markedly during the 12-hour period ending at 0000 UTC 10 February, when rapid surface deepening was in progress and heavy precipitation was spreading across New York and New England.

*500-mb geopotential height analyses*

- Prior to the major storm along the East Coast, a large cyclonic vortex drifted northward over eastern Canada on 8 February. No confluence was observed over the northeastern United States with only weak confluence over the Great Lakes as a weak surface high drifted eastward into New England on 8 February. The absence of confluence over the northeastern United States represents a

# 850 MB

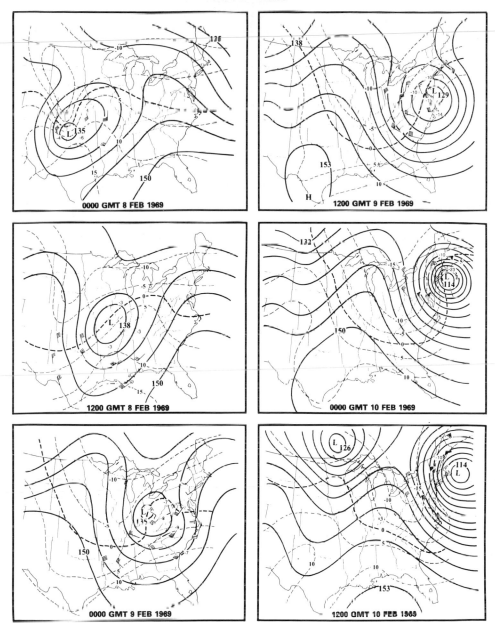

FIG. 76.    Twelve-hourly 850-mb analyses for 0000 UTC 8 February 1969 through 1200 UTC 10 February 1969. See Fig. 32 for details.

# 500 MB HEIGHTS AND UPPER-LEVEL WINDS

FIG. 77.   Twelve-hourly analyses of 500-mb geopotential heights and upper-level wind fields for 0000 UTC 8 February 1969 through 1200 UTC 10 February 1969. See Fig. 33 for details.

marked deviation from the precyclogenetic conditions of the other major snowstorms.

- The trough associated with the developing storm was located immediately downstream of an amplifying ridge in the northwestern United States and southwestern Canada. The trough propagated eastward from the central Plains as the primary cyclone moved into the Ohio Valley on 8 February, reaching the Middle Atlantic coast as the secondary cyclone developed along the East Coast on 9 February.

- A distinct vorticity maximum was evident at the base of the trough, where cyclonic curvature and shear were maximized. The primary and secondary cyclogenesis both occurred downstream of this vorticity maximum beneath the region of large vorticity advections.
- Between 0000 UTC 8 February and 0000 UTC 9 February, there was little change in the configuration of the geopotential height field that defined the trough as it propagated toward the East Coast. Between 1200 UTC 9 February and 1200 UTC 10 February, however, geopotential height values within the trough dropped significantly, with a closed 500-mb low developing by 0000 UTC 10 February. These changes occurred during the period of rapid surface intensification.
- Diffluence was observed downwind of the trough at all times. The rapid development of the secondary cyclone between 0000 and 1200 UTC 9 February coincided with the passage of the diffluent height contours across the coastline.
- The amplitude of the trough over the northeastern United States and the downstream ridge offshore increased, especially at 0000 UTC 10 February, as the ridge surged northward into eastern Canada during rapid cyclogenesis. The half-wavelength decreased between the trough and the downstream ridge off the East Coast, especially by 1200 UTC 9 February as precipitation spread across the Northeast and the cyclone began to deepen rapidly.

*Upper-level wind analyses*

- A subtropical jet was observed at 200 mb downstream of the trough which was approaching the East Coast between 0000 UTC 8 February and 0000 UTC 9 February. The subtropical jet was similar to those observed in other cases, although this jet did not amplify as it propagated across the southern United States.
- A separate jet upstream of the trough over the southern Plains appeared to propagate with the trough on 8 February, then merged with the subtropical jet along the Southeast coast by 1200 UTC 9 February. Wind speeds tended to increase near and upstream of the trough axis between 0000 UTC 8 February and 0000 UTC 9 February, prior to the East Coast surface development.
- The secondary surface low and its associated precipitation shield developed in the diffluent exit region of this jet system as it neared the East Coast after 0000 UTC 9 February.

## 7.13   22–28 FEBRUARY 1969

*General remarks*

- This was an unusual storm due to its slow movement, long duration, moderate intensity, erratic intensification, lack of large thermal contrast at the surface, and chaotic upper-level geopotential height patterns. The storm produced excessive amounts of snow across New England with accumulations of greater than 75 cm across large sections of eastern Massachusetts, New Hampshire, and Maine.

- Regions with snow accumulations exceeding 25 cm: central and northern Connecticut and Rhode Island, central and eastern Massachusetts (except Cape Cod and the Islands), eastern Vermont, New Hampshire, and most of Maine.
- Regions with snow accumulations exceeding 50 cm: eastern Massachusetts, northeastern Vermont, New Hampshire, and most of Maine.
- Urban center snowfall amounts:

  | | |
  |---|---|
  | Philadelphia, PA | 1.9″ (5 cm) |
  | New York, NY–La Guardia Airport | 1.7″ (4 cm) |
  | Boston, MA | 26.3″ (67 cm) |

- Other selected snowfall amounts:

  | | |
  |---|---|
  | Mt. Washington, NH | 97.8″ (248 cm) |
  | Pinkham Notch, NH | 77.3″ (196 cm) |
  | Long Falls Dam, ME | 56.0″ (142 cm) |
  | Old Town, ME | 43.6″ (111 cm) |
  | Rockport, MA | 39.0″ (99 cm) |
  | Portland, ME | 26.9″ (68 cm) |

*The surface chart*

- A large anticyclone was anchored over Hudson Bay prior to and during cyclogenesis (1040 mb). The air immediately along the East Coast during the storm period was maritime in origin, however, with temperatures generally between −5°C and 5°C.
- A ridge of high pressure extending down the East Coast prior to cyclogenesis reflected the cold air damming on 22 and 23 February.
- The development of a primary surface low was ill defined. An inverted trough initially in the Ohio Valley slowly evolved into a weak low center which meandered northeastward over the eastern Great Lakes on 23–24 February. This system exhibited no frontal structure and was attended by only light precipitation.
- A second inverted trough and coastal front developed along the Southeast coast by 1200 UTC 23 February. A weak cyclone then formed along the coastal front. The surface low deepened slowly [−4 mb (12 h)$^{-1}$] as it propagated from near the Virginia capes to a position east of New Jersey between 0000 and 1200 UTC 24 February. The low then intensified more rapidly in the following 12-hour period off the southern New England coast as its central pressure fell 10 mb to 994 mb. Heavy snowfall developed over southeastern New England during this period. Little deepening occurred thereafter. The surface low drifted very slowly after 1200 UTC 24 February, moving only 250 km in 24 hours.
- The cyclone meandered near the southeastern New England coast from 25–28 February, prolonging the heavy snowfall in eastern New England.
- Sea-level pressure gradients to the north and northeast of the surface low were

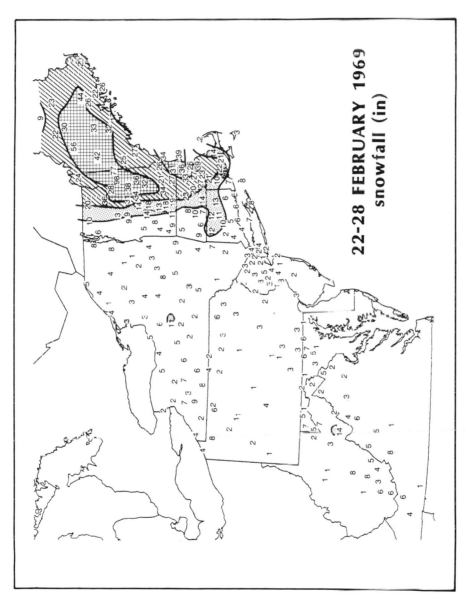

FIG. 78.   Snowfall (in) for 22–28 February 1969. See Fig. 30 for details.

# SURFACE

FIG. 79.    Twelve-hourly surface weather analyses for 1200 UTC 22 February 1969 through 1200 UTC 26 February 1969. See Fig. 31 for details.

# SURFACE (CONT.)

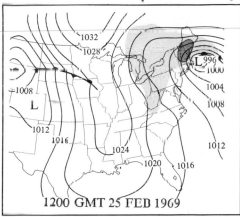

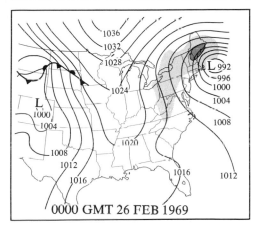

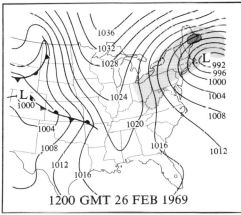

FIG. 79.   (*Continued*)

moderately intense, with winds approaching gale force only along the New England coast.

*The 850-mb chart*

- Cold air advection associated with northwesterly flow was not observed at 850 mb over the northeastern United States prior to this storm.
- Two separate 850-mb low centers were observed in this case. The first and weaker low moved northward from Kentucky to Michigan on 23 February. This system was associated with the inverted trough, surface low, and light precipitation over the Ohio Valley. The low was located in a weak and poorly-defined 850-mb temperature gradient. It propagated far to the north and west of the 0°C isotherm and did not deepen appreciably.
- A second 850-mb low formed by 1200 UTC 23 February along a band of somewhat enhanced temperature gradients near the Carolina coast. This center was associated with the sea-level cyclone that moved northeastward along the Atlantic seaboard. The 850-mb low deepened slowly at first, as it drifted up the East Coast. More vigorous deepening ensued after 1200 UTC 24 February, coinciding with the more rapid development of the surface low.
- The initial development of the 850-mb low was not marked by the strong cold air advection west of the center that characterized many of the other cases. A weak S-shaped isotherm pattern developed as warm air advection increased along the Northeast coast during 24 February. The low initially formed on the warm side of the 0°C isotherm, but was located closer to the 0°C line during cyclogenesis near New England late on 24 February and on 25 February.
- The cyclone evolved in the most diffuse 850-mb temperature field of the survey. However, a very narrow, yet significant, 850-mb temperature gradient formed over New England by 1200 UTC 24 February as the cyclone was beginning to intensify and heavy snow was falling in southeastern New England. Southeasterly to easterly winds and their associated moisture transports strengthened to the north of the coastal low center on 24 and 25 February as the system intensified and heavy snow fell across eastern New England.

*500-mb geopotential height analyses*

- The precyclogenetic geopotential height field at 500 mb did not exhibit the coherent trough/ridge patterns found in the other cases. An unusually large number of trough and ridge features can be identified across North America on 22–23 February. Troughs over the south-central United States and the southwestern United States at 1200 UTC 22 February later influenced the growth of the East Coast storm. The 500-mb chart also showed a trough off the West Coast, closed cyclonic circulations north of Montana and near Bermuda, and a ridge along the East Coast. A trough in southeastern Canada produced confluent flow, which was associated with the large surface anticyclone located over eastern Canada on 23 February.
- A trough in the Ohio Valley was associated with the inverted surface trough, weak cyclone, and light precipitation across the eastern United States on 23

# 850 MB

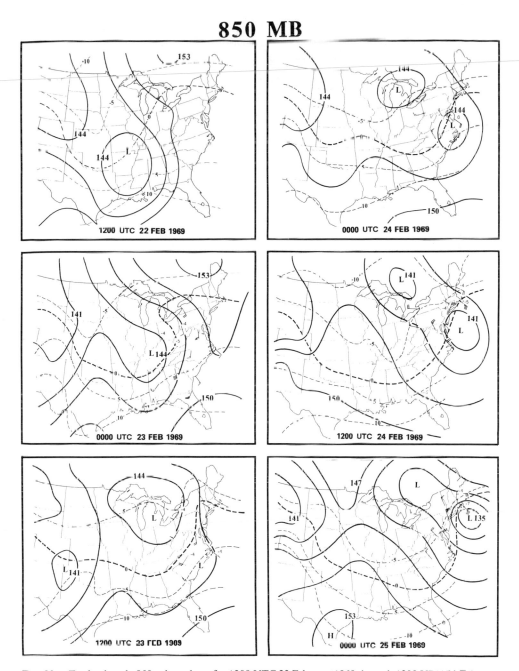

FIG. 80.    Twelve-hourly 850-mb analyses for 1200 UTC 22 February 1969 through 1200 UTC 26 February 1969. See Fig. 32 for details.

# 850 MB (CONT.)

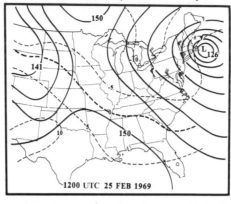

1200 UTC 25 FEB 1969

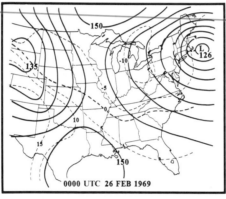

0000 UTC 26 FEB 1969

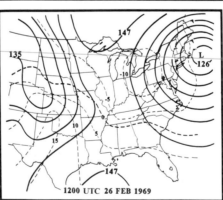

1200 UTC 26 FEB 1969

FIG. 80.   (*Continued*)

February. A second trough in the southwestern United States propagated to the East Coast by 24 February. It appeared to merge with the Ohio Valley trough, and may have influenced the more rapid development of the storm near New England on 24 February.

- Three separate cyclonic vorticity maxima were observed in connection with the two troughs over the central and eastern United States between 1200 UTC 23 February and 1200 UTC 24 February. The cyclonic vorticity maximum initially associated with the weak Ohio Valley trough first tracked northward across Michigan, then turned southeastward toward the Middle Atlantic coast by 26 February. Another vorticity maximum developed in the southeastern United States on 23 February in a region of increasing cyclonic wind shear associated with an intensifying jet streak upwind of the developing coastal cyclone. This feature propagated northeastward with the coastal low. The third vorticity maximum was associated with the separate trough/jet system that crossed the southern United States between 22 and 25 February.
- After 1200 UTC 24 February, the chaotic regime of small amplitude, high wavenumber features evolved into a simpler, more organized lower wavenumber pattern with troughs and ridges of increasing amplitude. A major trough remained off the West Coast with a ridge over central North America and a deep, closed cyclonic vortex along the Atlantic seaboard, reflecting the development of the surface cyclone near the East Coast.
- A pattern of weak diffluence was evident as early as 1200 UTC 23 February. A more pronounced pattern of diffluence was observed by 0000 and 1200 UTC 25 February, *following* the only 12-hour period in which the surface low intensified rapidly.
- Over the 36-hour period between 0000 UTC 23 February and 1200 UTC 24 February, during which the surface low deepened and moved toward New England, the amplitude between the troughs along the coast and the downstream ridge appeared to *decrease,* a deviation from the other cases examined. The half-wavelength, however, also decreased during this time.
- The passage of a trough across the southeastern United States on 24–25 February was followed by a general increase in amplitude as a longer-wavelength trough/ridge pattern and a probable blocking situation became established over the United States. The emergence of the low-wavenumber regime was associated with the slow movement of the storm off the coast for the following 3 to 4 days.

*Upper-level wind analyses*

- A jet marked by wind speeds of less than 50 m s$^{-1}$ at 300 mb was located in the confluent region south of a small cutoff low over eastern Canada on 23 February. This jet probably influenced the large surface anticyclone over eastern Canada, although this pattern was relatively weak and poorly defined as compared to the other cases.
- Although the 500-mb geopotential height fields were chaotic, clearly-defined upper-level jet streaks extended from northern Mexico to the Southeast coast from 23 to 25 February. These features were observed over the region of largest height gradients along the periphery of the chaotic trough pattern. Wind speeds generally exceeded 80 m s$^{-1}$ at both 200 and 300 mb for the entire observing period. The cyclonic wind shear associated with each jet streak imbedded within this flow regime contributed to the separate vorticity

# 500-MB HEIGHTS AND UPPER-LEVEL WINDS

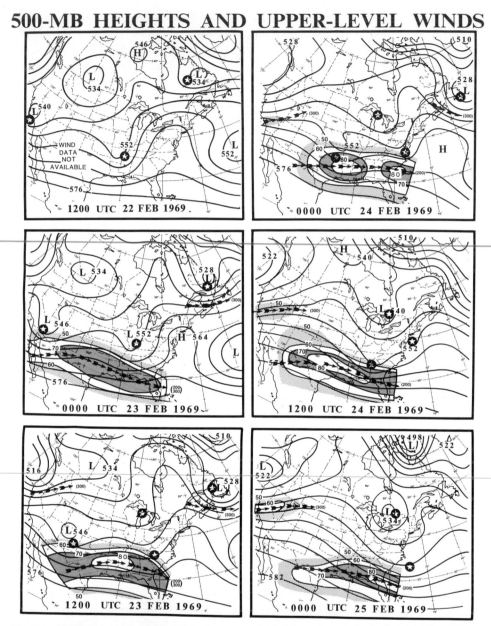

FIG. 81. Twelve-hourly analyses of 500-mb geopotential heights and upper-level wind fields for 1200 UTC 22 February 1969 through 1200 UTC 26 February 1969. See Fig. 33 for details.

maxima observed over the southern United States prior to cyclogenesis. The evolution of the various jet streaks and troughs into one coherent, negatively tilted system by 1200 UTC 25 February is difficult to describe for this case, given the lack of data off the East Coast.

## 500 MB HEIGHTS (CONT.)

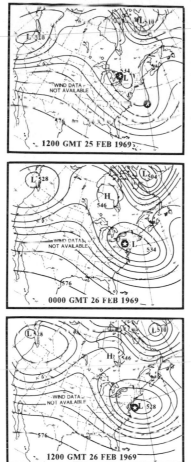

FIG. 81. (*Continued*)

## 7.14   25–28 DECEMBER 1969

*General remarks*

- This storm was a near-miss for the large cities of the northeastern United States as heavy snow turned to rain (and back to snow in many areas). The heaviest snow fell immediately north and west of the coastal plain. This system is one of the heaviest snowstorms on record for eastern and northern New York. Accumulations of greater than 50 cm covered a wide area of central and eastern New York into northwestern New England.
- Regions with snow accumulations exceeding 25 cm: western Virginia, western Maryland, portions of West Virginia, central and eastern Pennsylvania (except the extreme southeast), central and eastern New York (except the coast), Con-

necticut (except the coast), central and western Massachusetts, Vermont, New Hampshire, and western Maine.

- Regions with snow accumulations exceeding 50 cm: northeastern Pennsylvania, much of eastern New York, and Vermont.
- Urban center snowfall amounts:

| | |
|---|---|
| Washington, D.C.–Dulles Airport | 12.1″ (31 cm) |
| Baltimore, MD | 6.1″ (15 cm) |
| Philadelphia, PA | 5.2″ (13 cm) |
| New York, NY–La Guardia Airport | 7.4″ (19 cm) |
| Boston, MA | 4.2″ (11 cm) |

- Other selected snowfall amounts:

| | |
|---|---|
| Burlington, VT | 29.8″ (76 cm) |
| Albany, NY | 26.4″ (67 cm) |
| Binghamton, NY | 21.9″ (56 cm) |
| Williamsport, PA | 17.2″ (44 cm) |
| Roanoke, VA | 16.4″ (42 cm) |
| Hartford, CT | 15.2″ (39 cm) |
| Harrisburg, PA | 12.9″ (33 cm) |

*The surface chart*

- An anticyclone (1033 mb) accompanied by cold surface temperatures was centered north of New England prior to 26 December. The high center drifted slowly northeastward as cyclogenesis commenced along the coast.
- Cold air damming and coastal frontogenesis were prominent along the East Coast on 25 December and along the New England coast on 26 December.
- The surface low developed over Texas, well to the south of an occluding cyclone near the Minnesota–Canada border on 24–25 December. The low crossed the Gulf Coast states, then moved northeastward along the coastal front near the East Coast on 25–26 December. Precipitation spread rapidly northeastward with heavy snow falling along the Appalachian Mountains by late on 25 December. The storm track's close proximity to the coastline combined with the location of the surface anticyclone over extreme eastern Canada to draw warmer air inland, resulting in snow changing to rain over coastal areas. Central New England was plagued by a severe ice storm.
- The surface low moved rapidly northeastward along the East Coast between 0000 UTC and 1200 UTC 26 December, then slowed considerably as it reached the New Jersey shore. It then moved very slowly along the New England coast for the following 24 to 36 hours, advancing at only 5–10 m s$^{-1}$.
- Rapid intensification occurred during two periods. Between 0000 and 1200 UTC 26 December, the storm's central pressure fell 12 mb in 12 hours as it moved from Georgia to New Jersey. During this period, heavy precipitation spread quickly across the Middle Atlantic states into New England. The following 12 hours were marked by little intensification as the surface low moved very slowly from New Jersey to just east of Long Island. The period between 0000 and 1200 UTC 27 December brought the second surge of rapid deepening, as the central pressure fell another 12 mb to 976 mb, as the cyclone

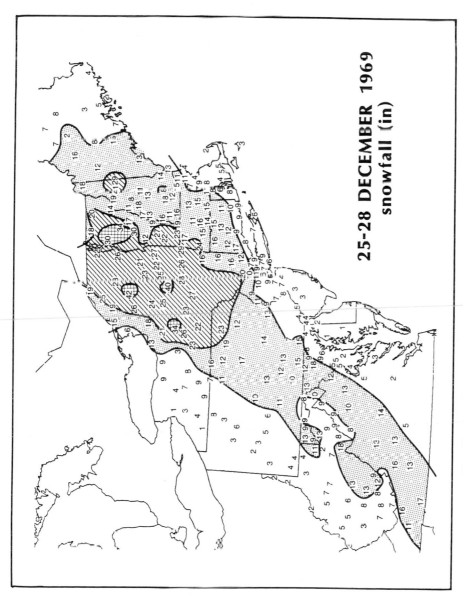

FIG. 82.   Snowfall (in) for 25–28 December 1969. See Fig. 30 for details.

# SURFACE

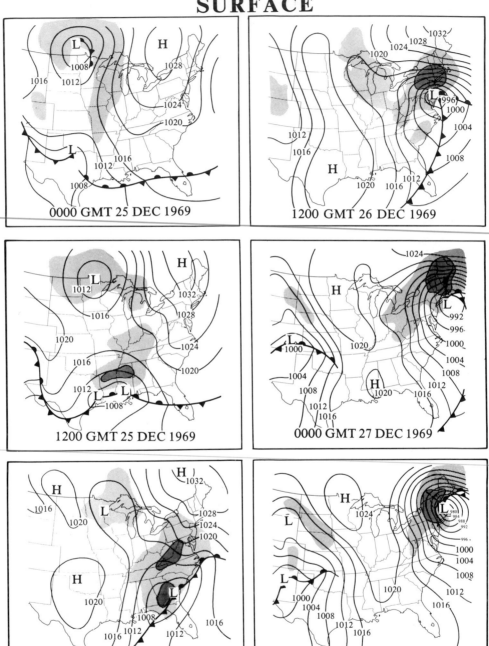

FIG. 83.   Twelve-hourly surface weather analyses for 0000 UTC 25 December 1969 through 1200 UTC 27 December 1969. See Fig. 31 for details.

drifted to the Massachusetts coast. Snowfall rates increased to the west of the storm center during this second period of intensification. The cyclone continued to drift slowly east-northeastward in the following 24 hours.

- The slow movement of the storm center after 1200 UTC 26 December contributed to the very heavy snowfall in northern New York and Vermont. Considerable warming also resulted over northern and eastern New England as persistent onshore winds raised temperatures well above freezing. The warm air from the Atlantic and a wedge of cold air remaining over western New England created a strong frontal boundary across central New England on 27 December.

*The 850-mb chart*

- Only weak northwesterly flow and cold air advection were observed across the northeastern United States on 24 December, prior to the snowstorm. The 850-mb temperatures ranged from 0°C over North Carolina to −15°C across northern New York and New England.

- An 850-mb low center was located near the Canadian border at 0000 UTC 25 December in association with the occluding surface low. The cyclonic circulation around this center produced warm air advection across the Great Lakes region through 0000 UTC 26 December.

- The 850-mb low center that was associated with the major storm drifted eastward from the south-central to the southeastern United States. The low center deepened very slowly on 25 December as the surface low crossed the Gulf coast. It then experienced two 12-hour periods of rapid deepening, concurrent with the intensification of the surface low. The first surge occurred between 0000 UTC and 1200 UTC 26 December [−90 m (12 h)$^{-1}$], with the second burst between 0000 UTC and 1200 UTC 27 December [−90 m (12 h)$^{-1}$]. During the 12-hour interval separating these periods, the 850-mb low center deepened at a rate of only −30 m (12 h)$^{-1}$.

- The 850-mb low developed along an east–west thermal ribbon extending across the southern United States. Cold air advection behind the low increased across Texas by 1200 UTC 25 December while strong warm air advection east of the low coincided with the rapid outbreak of moderate to heavy precipitation. These amplifying advections resulted in an S-shaped isotherm pattern along the coast by 1200 UTC 26 December. This pattern gradually became oriented north to south from the Carolinas to Maine on 27 December, as the low moved over Massachusetts and warm air flooded coastal New England, changing the snow to rain in that region.

- A southerly LLJ developed over Mississippi and Alabama by 1200 UTC 25 December and shifted to the Southeast coast by 0000 UTC 26 December, immediately prior to the first period of rapid deepening along the Atlantic coast. Coastal frontogenesis occurred and precipitation broke out across the southeastern states during this period. The jet formed as the rapidly advancing and deepening 850-mb low encountered a slower-moving ridge to its east. As a result, the geopotential height gradients increased along the Southeast coast.

# 850 MB

FIG. 84.    Twelve-hourly 850-mb analyses for 0000 UTC 25 December 1969 through 1200 UTC 27 December 1969. See Fig. 32 for details.

- An easterly jet developed to the north of the low by 1200 UTC 26 December, as the cyclone was undergoing its first period of rapid deepening. The southerly and easterly low-level jets enhanced the moisture transports into regions of heavy precipitation.

*500-mb geopotential height analyses*

- The period prior to 26 December was marked by nearly zonal flow across much of the United States. No high-amplitude features were present anywhere over North America on 24–25 December. A trough over eastern Canada did not contain the geopotential height minimum, closed cyclonic circulation, or intense vorticity gradients that marked many of the other cases. The flow over New England and the Great Lakes was not characterized by the degree of confluence seen in the precyclogenetic periods of many other cases. A weakly-defined pattern of confluent geopotential heights was found across southern Ontario and Quebec at 0000 and 1200 UTC 25 December, above the surface anticyclone. A better-defined pattern of confluence appeared across extreme eastern Canada at 1200 UTC 26 December and 0000 UTC 27 December as the surface high drifted toward Newfoundland.
- The cyclone's development was linked to the amplification of a trough that propagated eastward from the central United States after 0000 UTC 25 December. A closed cyclonic circulation formed at 500 mb along the East Coast by 1200 UTC 27 December. The amplifying trough seemed to pivot about a small cyclonic vortex over the Great Lakes region on 26 December. This feature was associated with the dying surface system over the northern United States.
- The cyclonic vorticity maximum and vorticity advection pattern were associated with the trough and jet streak system which propagated eastward from the south-central United States.
- As the 500-mb trough deepened over the southeastern United States by 1200 UTC 26 December, it became oriented from northwest to southeast (a negative tilt), with a diffluent pattern appearing downwind of the trough axis. These changes in trough structure occurred as the surface low moved rapidly toward New England and went through its first period of significant deepening. By 1200 UTC 27 December, diffluence was observed over New England between the trough and the downstream ridge. This coincided with the second period of rapid deepening, as the surface low moved very slowly along the New England coast.
- The cyclogenetic period from 26 to 27 December was characterized by a general increase of amplitude across the United States. A large increase in amplitude occurred between the trough over the eastern United States and the downstream ridge over New England and southeastern Canada on 26 December.
- The half-wavelength of the trough and downstream ridge shortened during cyclogenesis, especially between 0000 UTC 26 December and 0000 UTC 27 December.

# 500-MB HEIGHTS AND UPPER-LEVEL WINDS

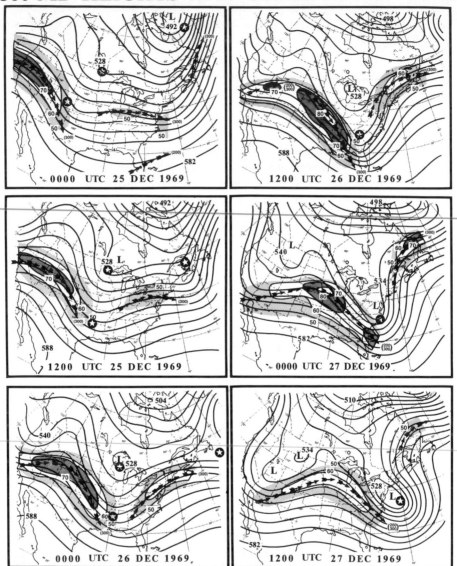

FIG. 85.    Twelve-hourly analyses of 500-mb geopotential heights and upper-level wind fields for 0000 UTC 25 December 1969 through 1200 UTC 27 December 1969. See Fig. 33 for details.

*Upper-level wind analyses*

- Unlike nearly all the other cases, the jet maxima were primarily located in the ridge crest *upstream* of the trough. Winds in excess of 70 m s$^{-1}$ were found at the 300-mb level throughout the study period.
- The jet streak crossing the West Coast on 25 December propagated toward the base of the amplifying trough nearing the East Coast by 1200 UTC 26

December. By that time, the surface low was already located over southern New Jersey, well to the north of this jet streak.

- The initial development and rapid northeast motion of the surface low after 0000 UTC 26 December occurred in the entrance region of a separate jet streak in the downstream ridge over New England. This jet amplified from 60 m s$^{-1}$ to 70 m s$^{-1}$ during 26 December, as the ridge crest became better defined over the northeastern United States and heavy precipitation spread up the East Coast.

## 7.15    18–20 FEBRUARY 1972

*General remarks*

- This storm was another near-miss for the major cities as the heaviest snow fell immediately to their west and north. This storm was one of the few to pose a threat of heavy snow in the Northeast urban corridor during the early and middle 1970s. Strong easterly winds and rough seas caused significant damage along the Middle Atlantic and New England coasts.
- Regions with snow accumulations exceeding 25 cm: eastern West Virginia, northern Virginia, central and northern Maryland, Pennsylvania (except the west and southeast), northwestern New Jersey, New York (except the extreme west and coastal regions), central and northern Connecticut, Massachusetts (except the southeast), Maine, and parts of New Hampshire and Vermont.
- Regions with snow accumulations exceeding 50 cm: north-central Pennsylvania, central New York, and parts of West Virginia.
- Urban center snowfall amounts:

| | |
|---|---|
| Washington, D.C.–Dulles Airport | 10.0" (25 cm) |
| Baltimore, MD | 3.2" (8 cm) |
| Philadelphia, PA | 3.7" (9 cm) |
| New York, NY–La Guardia Airport | 6.3" (16 cm) |
| Boston, MA | 6.3" (16 cm) |

- Other selected snowfall amounts:

| | |
|---|---|
| Binghamton, NY | 24.4" (62 cm) |
| Williamsport, PA | 22.8" (58 cm) |
| Syracuse, NY | 20.0" (51 cm) |
| Portland, ME | 15.4" (39 cm) |
| Worcester, MA | 15.2" (39 cm) |
| Harrisburg, PA | 13.0" (33 cm) |
| Albany, NY | 12.1" (31 cm) |

*The surface chart*

- An anticyclone (1035 mb) was located to the northeast of New England on 18 February, prior to East Coast cyclogenesis. Its eastward location, in concert with the surface low's track close to the shoreline, produced an east-to-northeasterly flow along the coast. The resultant warming ensured that a significant

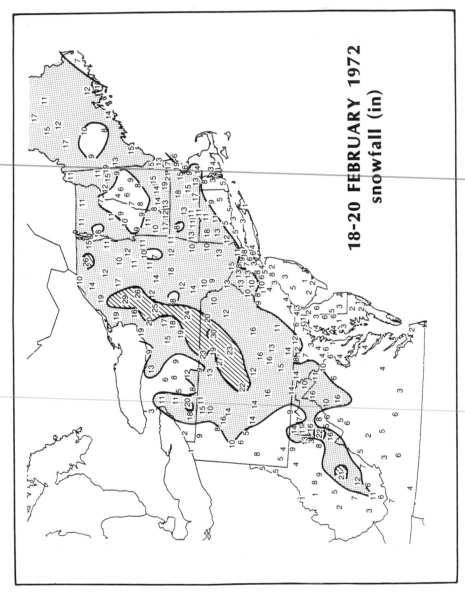

FIG. 86.    Snowfall (in) for 18–20 February 1972. See Fig. 30 for details.

portion of the precipitation would fall as rain within the major metropolitan areas.

- Cold air damming and coastal frontogenesis were observed along the Southeast coast on 18 February, prior to the coastal cyclogenesis.
- The primary low was a slow-moving, intense storm over the upper Midwest. It filled 10 mb in the 24 hours between 0000 UTC 18 February and 0000 UTC 19 February as it moved over the Great Lakes. Despite the obvious dissipation of the primary low, it remained evident for 27 hours after the onset of secondary cyclogenesis.
- A secondary low developed over southern Georgia at 1200 UTC 18 February, 1600 km south of the primary center. This separation represents the largest distance between primary and secondary low centers in the sample. The secondary low deepened rapidly after 0300 UTC 19 February, with the central sea-level pressure falling 25 mb in 12 hours, reaching 975 mb over New Jersey by 1500 UTC 19 February.
- The secondary low propagated northeastward at an average rate of 15 m s$^{-1}$ along the Southeast coast. The forward motion slowed along the Middle Atlantic and New England shores on 19 February. By 1800 UTC 19 February, the cyclone developed dual centers as it moved slowly and erratically off the New Jersey and New England coasts.
- An intense pressure gradient developed to the northeast of the rapidly developing storm on 19 February and was associated with very strong winds, producing a strong easterly fetch of air toward the Middle Atlantic and southern New England shorelines. Convergent winds associated with an inverted trough extending northward from the secondary low may have been instrumental in the development of heavy snowfall over Pennsylvania, Maryland, and Virginia on 19 February.

*The 850-mb chart*

- The 24-hour period prior to cyclogenesis on the East Coast did not feature the northwesterly flow and cold air advection over the Great Lakes and New England that characterized most of the other cases. A small cyclonic system moving eastward over the western Atlantic on 17 February contributed to a pattern of weak cold air advection in the Middle Atlantic states. The 850-mb 0°C isotherm was displaced to South Carolina by 0000 UTC 18 February.
- An intense 850-mb low center drifted slowly eastward over the upper Great Lakes region through 0000 UTC 19 February. This system was located well to the north of the 0°C isotherm. It weakened rapidly in the 24-hour period after 1200 UTC 18 February as the largest temperature gradient progressed southward away from the low center.
- The circular low center over the upper Great Lakes became elongated on a north–south axis during 18 February as cold air advection west of the trough drove the 0°C isotherm southward to the Gulf coast. A new center developed over the Carolinas by 0000 UTC 19 February, near the 0°C isotherm. The Great Lakes low center disappeared by 1200 UTC 19 February as the secondary 850-mb circulation intensified rapidly along the East Coast. A northerly jet

# SURFACE

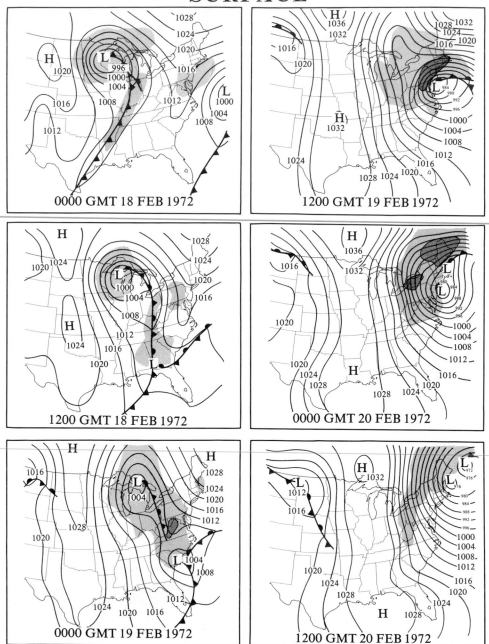

FIG. 87.   Twelve-hourly surface weather analyses for 0000 UTC 18 February 1972 through 1200 UTC 20 February 1972. See Fig. 31 for details.

# 850 MB

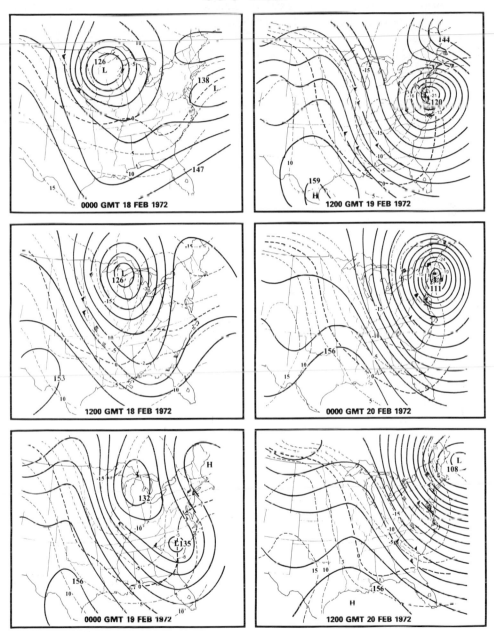

FIG. 88. Twelve-hourly 850-mb analyses for 0000 UTC 18 February 1972 through 1200 UTC 20 February 1972. See Fig. 32 for details.

with wind speeds exceeding 25 m s$^{-1}$ was established west of the low centers prior to secondary cyclogenesis, with cold air advection occurring throughout the south-central and southeastern United States.

- The secondary 850-mb low deepened 150 m in the 12-hour period ending at 1200 UTC 19 February and 90 m in the following 12-hour period.
- The secondary development began concurrent with an increase in warm air advection along the East Coast by 0000 UTC 19 February. The 850-mb low developed near the inflection point of a developing S-shaped isotherm pattern. The temperature gradient intensified along the East Coast by 1200 UTC 19 February during a period when the surface and 850-mb lows deepened rapidly.
- An intense southeasterly-to-easterly jet formed to the north of the rapidly deepening low between 0000 and 1200 UTC 19 February. The jet was directed toward and coincided with the outbreak of heavy snow and rain in the Middle Atlantic states and New England. This suggests that the increasing moisture inflow associated with the intensifying low-level circulation was an important factor in the expanding area of heavy precipitation on 19 February.

*500-mb geopotential height analyses*

- A trough over eastern Canada was associated with confluence across the maritime provinces on 18–19 February. This confluence region was associated with the surface anticyclone over eastern Canada that funneled cold air down the East Coast on 17 February. The Canadian trough and its associated confluence zone continued moving eastward over the Atlantic Ocean during the following two days.
- By 0000 UTC 19 February, the confluent area was located far to the north and east of New England. The withdrawal of the confluence zone from the northeastern United States was consistent with the retreat of cold air along the coast and the changeover from snow to rain which occurred in many places on 19 February.
- A slowly amplifying ridge was established over the western United States on 18 February. The trough that was later linked to the major East Coast storm amplified over the central United States prior to cyclogenesis. This trough continued to amplify through 1200 UTC 19 February as it propagated toward the East Coast.
- A 500-mb geopotential height minimum over the western Great Lakes on 18 February was nearly collocated with the weakening primary surface low. This feature disappeared by 1200 UTC 19 February as a new cyclonic vortex deepened rapidly over the eastern United States in association with the secondary cyclogenesis along the Middle Atlantic coast.
- The cyclone over the Great Lakes was associated with a cyclonic vorticity maximum that weakened on 18–20 February. The developing cyclone along the East Coast was associated with a separate increasing vorticity maximum that swept northeastward along the Southeast and Middle Atlantic coasts on 19 February. This vorticity maximum intensified as the curvature increased near the base of the amplifying trough and the cyclonic shear associated with

# 500 MB HEIGHTS AND UPPER-LEVEL WINDS

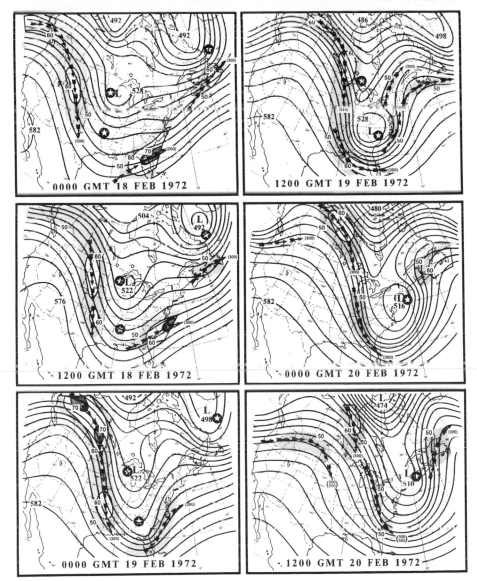

FIG. 89.  Twelve-hourly analyses of 500-mb geopotential heights and upper-level wind fields for 0000 UTC 18 February 1972 through 1200 UTC 20 February 1972. See Fig. 33 for details.

the jet strengthened, especially between 1200 UTC 18 February and 1200 UTC 19 February.

- The rapid development of the secondary cyclone occurred beneath an increasingly diffluent geopotential height pattern downwind of the trough between 0000 and 1200 UTC 19 February, as the trough axis became oriented from northwest to southeast (a negative tilt). Geopotential height gradients at the base of the trough increased at 0000 UTC 19 February, immediately prior to the rapid secondary cyclogenesis.

- The amplitude of the trough and downstream ridge increased between 0000 UTC 18 February and 1200 UTC 19 February. There was also a shortening of the half-wavelength between the trough and downstream ridge during the secondary development, especially between 0000 UTC 19 February and 0000 UTC 20 February.

*Upper-level wind analyses*

- A jet streak with maximum wind speeds in excess of 70 m s$^{-1}$ was located over the southeastern United States in advance of the cyclone on 18 February. This wind maximum was located downstream of the trough axis over the central United States and upstream of a minor trough moving out over the Atlantic Ocean.
- A series of upper-level wind maxima propagated from the ridge over the West Coast toward the base of the amplifying trough in the central United States. It was difficult to identify and follow any particular jet streak during this period.
- Winds at the base of the trough were difficult to analyze at the time of secondary cyclogenesis since the strongest winds were located over data-void regions off the East and Gulf coasts. At 1200 UTC 19 February, it appeared that a jet extended from the base of the trough in the southeastern United States toward the diffluent region downwind of the trough axis. The surface low deepened rapidly off the Middle Atlantic coast within the diffluent exit region of this jet system.

## 7.16    19–21 JANUARY 1978

*General remarks*

- For many of the urban centers that span the northeastern United States, this was the most debilitating snowstorm since 1969. This storm was the last in a series of three during a week that produced a variety of winter weather conditions across the Northeast. Along the coast, snowfall was underforecast since a predicted changeover from snow to rain either did not occur or took place after there had already been substantial accumulations. Boston, Massachusetts, set its 24-hour snowfall record with this storm, only to have it broken two weeks later. The storm was accompanied by wind gusts exceeding 20 m s$^{-1}$ from New Jersey to the New England coast.
- Regions with snow accumulations exceeding 25 cm: sections of West Virginia, western and northern Virginia and Maryland, much of Pennsylvania, central and northern New Jersey, and much of New York and New England.
- Regions with snow accumulations exceeding 50 cm: portions of West Virginia, western Maryland, central and northeastern New York, eastern Massachusetts, and northern Rhode Island.

- Urban center snowfall amounts:

| | |
|---|---|
| Washington, D.C.–Dulles Airport | 7.5" (19 cm) |
| Baltimore, MD | 5.6" (14 cm) |
| Philadelphia, PA | 13.2" (34 cm) |
| New York, NY–Kennedy Airport | 14.2" (36 cm) |
| Boston, MA | 21.5" (54 cm) |

- Other selected snowfall amounts:

| | |
|---|---|
| Syracuse, NY | 18.9" (48 cm) |
| Charleston, WV | 18.4" (47 cm) |
| Newark, NJ | 17.8" (45 cm) |
| Bridgeport, CT | 16.7" (42 cm) |
| Hartford, CT | 15.5" (39 cm) |
| Pittsburgh, PA | 14.0" (36 cm) |

*The surface chart*

- A surface low associated with the second in a series of three major snowstorms was located off the Maine coast at 0000 UTC 19 January. This storm was responsible for heavy snow accumulations across inland sections of the northeastern United States on 17–18 January. Rising sea-level pressures and colder air followed the passage of this storm as a large, cold anticyclone (1046 mb) over the northern Plains states on 19 January built eastward toward northern New England.
- On 19 January, cold air damming and an inverted sea-level pressure ridge developed east of the Appalachian Mountains. At the same time, an inverted trough and coastal front developed along the Carolina coast.
- The cyclone formed over the Gulf of Mexico on 18 January and moved northeastward across the eastern Gulf of Mexico and along the East Coast on 19–20 January. The low propagated at about 15–20 m s$^{-1}$ and tended to redevelop or "jump" northeastward along the coastal front near the Carolina coast between 0000 UTC and 1200 UTC 20 January.
- This storm exhibited rather erratic intensification. After forming in the Gulf of Mexico, the central pressure of the cyclone vacillated as it moved northeastward. It deepened rapidly for only a brief period off the New Jersey coast near 1800 UTC 20 January, at which time the center reached its lowest pressure of 995 mb.
- Although the minimum sea-level pressure was not particularly low relative to the other cases, the large areal extent of the cyclone's circulation and its interaction with the Canadian anticyclone to the north produced a widespread snowstorm accompanied by gale-force winds along the Atlantic coast. Winds increased along the Middle Atlantic coast on 20 January as the distance between the anticyclone and the advancing surface low decreased.
- The heaviest precipitation fell several hundred kilometers in advance of the surface low center, especially from 0000 through 1200 UTC 20 January.

*The 850 mb chart*

- West to northwesterly flow to the rear of the cyclone moving northeastward past the Canadian maritime provinces maintained low-level cold air advection

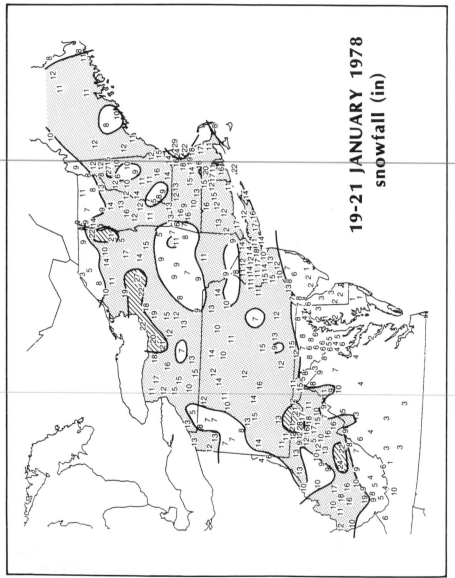

FIG. 90.    Snowfall (in) for 19–21 January 1978. See Fig. 30 for details.

# SURFACE

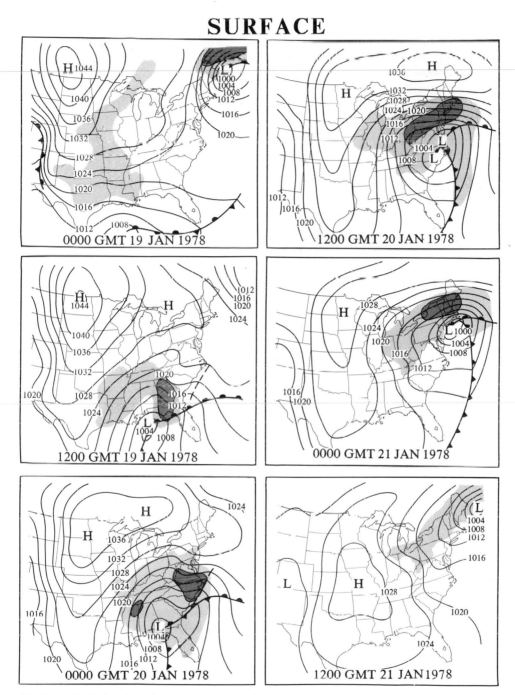

FIG. 91.  Twelve-hourly surface weather analyses for 0000 UTC 19 January 1978 through 1200 UTC 21 January 1978. See Fig. 31 for details.

across the northeastern United States through 0000 UTC 20 January. The cold air moved into New England and the Middle Atlantic states as an 850-mb low developed along the Gulf coast.

- This 850-mb low developed in a broad, southwest–northeast baroclinic zone extending across the southeastern United States early on 19 January. The cyclonic circulation moved northeastward along the isotherm ribbon and expanded in areal extent during 19 February, covering much of the eastern United States by 20 January.

- The 850-mb low deepened only slowly as it propagated from the Texas coast to southern New England. The greatest intensification [$-60$ m $(12$ h$)^{-1}$] occurred between 1200 UTC 19 January and 0000 UTC 20 January, although the surface low deepened only slightly during the same period. The small 850-mb deepening rates are consistent with those of the surface low for this system.

- As the cyclone propagated northeastward, the 850-mb temperature gradient increased in advance of the low, especially on 20 January, as heavy snow spread across the Middle Atlantic states and southern New England. The S-shaped isotherm pattern observed in many other storms was not as pronounced in this case since the cold air advection to the rear of the cyclone remained weak, especially after 0000 UTC 20 January. The low developed on the warm side of the 0°C isotherm on 18 January, but tracked closer to the 0°C isotherm by the time it reached Virginia on 20 January.

- A southerly LLJ developed across Florida by 1200 UTC 19 January, immediately upwind of an outbreak of moderate to heavy precipitation over the southeastern United States. Easterly winds strengthened to the north of the low center as it intensified slightly and moved from Louisiana to Virginia between 1200 UTC 19 January and 1200 UTC 20 January. The southerly LLJ at 1200 UTC 20 January and the easterly LLJ at 0000 UTC 21 January were both directed toward the region of heaviest precipitation.

*500-mb geopotential height analyses*

- As the surface cyclone was forming in the Gulf of Mexico on 18 January, two separate troughs encircled an upper-level vortex over northern Quebec. One trough was associated with the cyclone that had brought heavy snowfall to sections of the northeastern United States on 17–18 January and was located over Newfoundland by 1200 UTC 19 January. This trough was followed by a significant increase in geopotential heights over the western Atlantic on 19 January. The second trough extended from the vortex over northern Quebec to Nebraska at 0000 UTC 19 January and later split into two separate trough features. The southern part developed into a cyclonic circulation over the Midwestern states while the northern portion pulled away and propagated across eastern Canada on 19 January.

- The combination of rising geopotential heights off the East Coast and the passage of a trough across eastern Canada resulted in a strongly confluent height field near the United States–Canada border between 0000 and 1200 UTC 20 January. This pattern was observed as the surface anticyclone built

# 850 MB

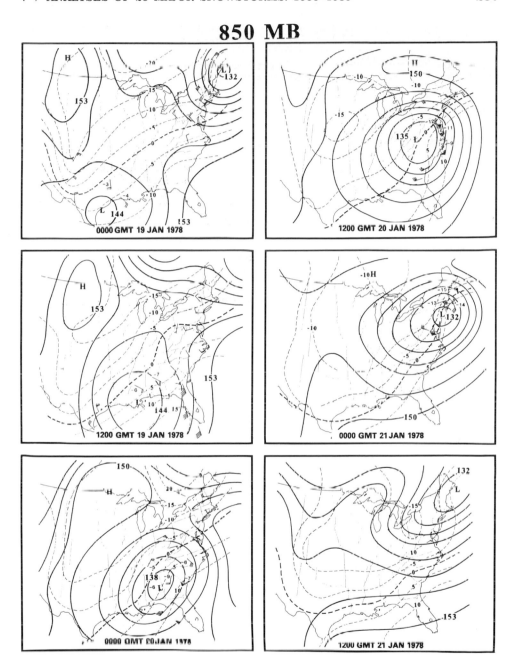

FIG. 92.   Twelve-hourly 850-mb analyses for 0000 UTC 19 January 1978 through 1200 UTC 21 January 1978. See Fig. 32 for details.

# 500 MB HEIGHTS AND UPPER-LEVEL WINDS

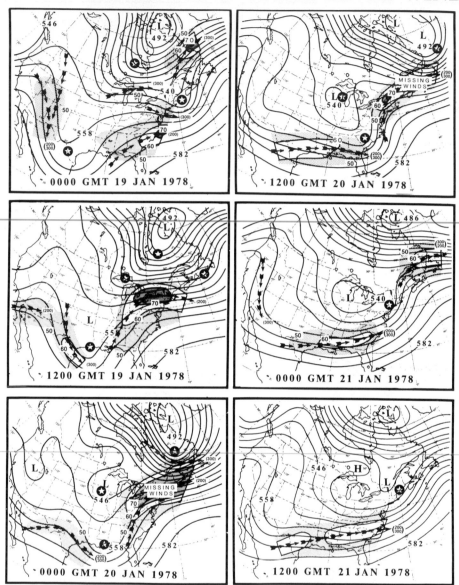

FIG. 93. Twelve-hourly analyses of 500-mb geopotential heights and upper-level wind fields for 0000 UTC 19 January 1978 through 1200 UTC 21 January 1978. See Fig. 33 for details.

eastward across southern Quebec and cold air damming developed east of the Appalachian Mountains.

- The trough that was associated with the Gulf cyclone propagated across the southern United States on 18 and 19 January, then moved northeastward along the East Coast on 20 January. This trough rotated about the trough developing over the Midwest. Both troughs were associated with well-defined cyclonic vorticity maxima that maintained their individual identities through 1200 UTC 20 January.

- A closed 500-mb circulation did not develop in association with the cyclone tracking northeastward to New England. The amplitude of the trough and downstream ridge along the Atlantic coast decreased slightly between 1200 UTC 19 January to 0000 UTC 21 January. This is only one of two cases in which the amplitude decreased.
- The trough axis became negatively tilted (oriented from northwest to southeast) and the half-wavelength between the trough and downstream ridge decreased significantly by 1200 UTC 20 January; but the flow appeared to become only slightly diffluent east of the trough axis between 1200 UTC 20 January and 1200 UTC 21 January. During this period, the surface low deepened by only 4 mb, although the heaviest snow was falling across the northeastern United States. The lack of a strongly diffluent height pattern may explain the small surface deepening rate observed in this storm.

*Upper-level wind analyses*

- A jet streak amplified from 50 m s$^{-1}$ to 70 m s$^{-1}$ in the ridge over New England by 1200 UTC 19 January within a region of strong confluence. The cold surface anticyclone ridged southeastward into New England and toward the Middle Atlantic states beneath this confluence zone.
- Missing wind reports over New England and southeastern Canada on 20 January may indicate that winds continued to increase within the jet streak. The precipitation area to the north and northeast of the storm center developed and expanded in the entrance region of this jet as it crossed New England between 1200 UTC 19 January and 1200 UTC 20 January.
- Separate jet streaks propagated toward the base of the trough over the Gulf coast on 19 and 20 January. By 0000 UTC 21 January, one well-defined jet streak was observed over the southeastern United States.
- The surface low was located on the anticyclonic side of the entrance region of the jet streak over the northeastern United States between 1200 UTC 19 January to 0000 UTC 21 January, and in the exit region of the separate jet streak propagating around the base of the trough in the southeastern United States during the same period. This is similar to many of the other cases examined here.

*Infrared satellite imagery sequence*

- The initial cyclone development in the Gulf of Mexico was characterized by a cold cloud top core, as indicated by the dark shade enhancement south of Louisiana at 0000 UTC 19 January. This area, which may have contained intense convective elements, moved over Florida by 1200 UTC 19 January.
- The rapid northeastward expansion of the high cloud shield on 19 and 20 January extended over 1000 km north of the surface low-pressure center. This expansion occurred in the entrance region of the amplifying polar jet over

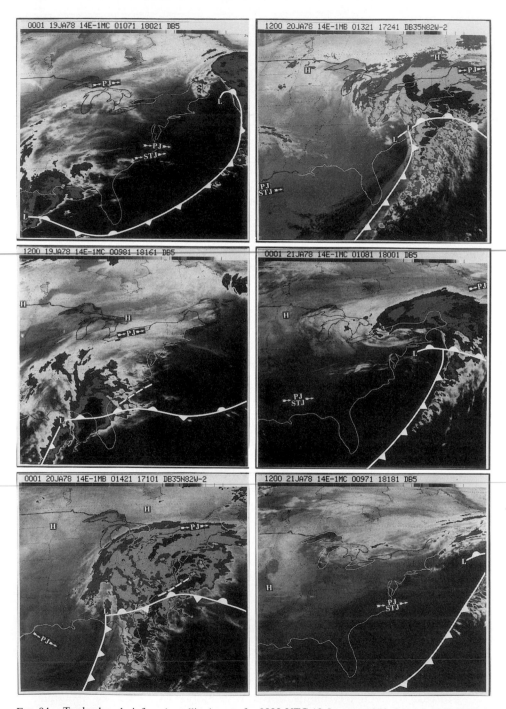

FIG. 94.   Twelve-hourly infrared satellite images for 0000 UTC 19 January 1978 through 1200 UTC 21 January 1978. Surface fronts and low- and high-pressure centers are shown, along with the locations of upper-level polar (PJ) and subtropical (STJ) jet streaks.

the northeastern United States. The sharp northern edge of the high cloud mass corresponded with the axis of the confluent upper-level flow just north of New England.

- A clear tongue marked by a region free of upper-level clouds developed along the Gulf coast at 0000 UTC 20 January and moved into North Carolina by 1200 UTC 20 January. The clear wedge corresponded to the axis of the jet located downwind of the upper-level trough (see 500-mb heights and upper-level winds chart). The surface low was located very close to the eastern edge of the dry slot or along the western edge of the cloud area.
- The upper cloud mass developed a distinct comma shape by 0000 UTC 21 January as the northern part of the cloud area pivoted northward while the comma tail swept to the east.
- The heaviest precipitation was located within the anticyclonically curved cloud region marking the warm conveyor belt (see section 5.5) through 1200 UTC 20 January, but then shifted toward the southern and western edges of the comma head, or cold conveyor belt (see section 5.5), as the low moved toward New England. It is difficult to distinguish the separate airstreams from the infrared images.

Note: The "gray scale" that borders the top of each image provides a means of using black, white, and shades of gray to distinguish intervals of cloud-top temperature. The gray scale at 0000 and 1200 UTC 20 January differs from the scale at the other times.

## 7.17  5–7 FEBRUARY 1978

*General remarks*

- Hurricane-force winds and record-breaking snowfall made this storm one of the more intense to occur this century across parts of the northeastern United States. A small area of 50 or more inches of snowfall was reported in Northern Rhode Island (not depicted in Figure 95). The cyclone was forecast remarkably well several days in advance by operational numerical forecast models (Brown and Olson 1978). Despite these accurate predictions, many people were stranded on the roads in the New York City area, because the onset of heavy snow occurred slightly later than predicted during the Monday morning rush hour. People were generally skeptical of the warnings issued by operational weather forecasters following a series of inaccurate forecasts of winter weather during the preceding month. The most severely affected regions were Long Island, Connecticut, Rhode Island, and Massachusetts, where businesses and schools were shut down for a week or more. This storm is a remarkable example of sudden and rapid cyclonic development associated with a major trough amplification at 500 mb.
- Regions with snow accumulations exceeding 25 cm: eastern Maryland, Delaware, eastern Pennsylvania, New Jersey, southeastern, northeastern and portions of western New York, Connecticut, Rhode Island, Massachusetts, central and southern Vermont, New Hampshire, and Maine.
- Regions with snow accumulations exceeding 50 cm: sections of northeastern

Pennsylvania, northern New Jersey, western and southeastern New York, Connecticut, Rhode Island, Massachusetts, southern Vermont, and parts of New Hampshire and Maine.

- Urban center snowfall amounts:

| | |
|---|---|
| Washington, D.C.–National Airport | 2.2″ (6 cm) |
| Baltimore, MD | 9.1″ (23 cm) |
| Philadelphia, PA | 14.1″ (36 cm) |
| New York, NY–Central Park | 17.7″ (45 cm) |
| Boston, MA | 27.1″ (69 cm) |

- Other selected snowfall amounts:

| | |
|---|---|
| Woonsocket, RI | 38.0″ (97 cm) |
| Rockport, MA | 32.5″ (83 cm) |
| Providence, RI | 28.6″ (73 cm) |
| Rochester, NY | 25.8″ (66 cm) |
| Riverhead, NY | 25.0″ (64 cm) |
| Worcester, MA | 20.2″ (51 cm) |
| Hartford, CT | 16.9″ (43 cm) |
| Trenton, NJ | 16.1″ (41 cm) |
| Wilmington, DE | 14.5″ (37 cm) |

*The surface chart*

- The cyclone developed in a regime of unusually high sea-level pressure and very cold temperatures. A 1055-mb anticyclone drifted across central Canada prior to and during the storm, with two high-pressure ridges dominating the precyclogenetic period on 5–6 February. One ridge axis extended southward through the center of the country and the other extended from New England to the southeastern United States.
- Weak cold air damming was indicated by the inverted high-pressure ridge along the East Coast on 5 February. Despite the wedge of very cold air along the Atlantic seaboard, no coastal frontogenesis occurred in this case.
- A weak 1025-mb low crossing the Great Lakes states on 5 February produced only light snows and provided few clues that major cyclogenesis would occur the following day. This low dissipated early on 6 February following the development of a secondary cyclone east of North Carolina.
- The secondary low developed in a data-void region, but appeared to deepen at rates exceeding $-3$ mb (3 h)$^{-1}$ between 0600 UTC 6 February and 0000 UTC 7 February as it moved northward toward Long Island. Since the cyclone developed in an environment dominated by high pressure, it deepened to only 984 mb at its peak intensity. This ranks only 14th out of the 20 storms in this sample in terms of the lowest central sea-level pressure, but the pressure gradient to the north and west of the low center was among the largest in the sample, producing gale- to hurricane-force northeasterly winds which drifted the heavy snow, reducing visibilities to near zero over a large area.
- The surface low propagated at approximately 10–12 m s$^{-1}$ as it tracked northward from a position well off the North Carolina coast to just south of Long Island on 6 February. The cyclone then may have performed a small coun-

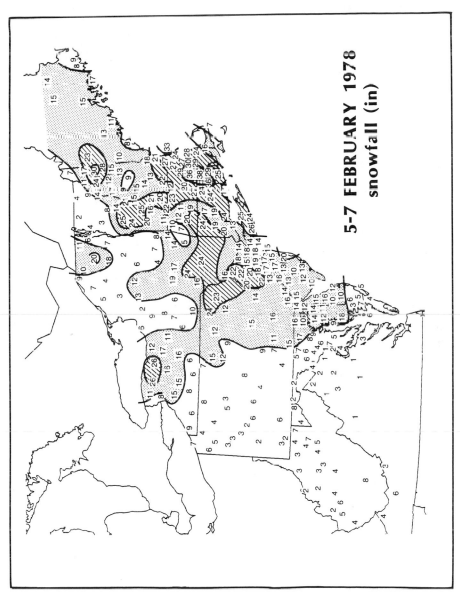

FIG. 95.   Snowfall (in) for 5-7 February 1978. See Fig. 30 for details.

# SURFACE

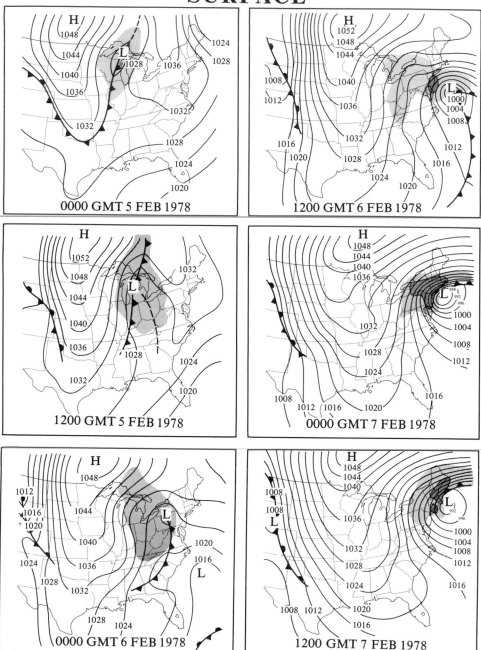

FIG. 96.   Twelve-hourly surface weather analyses for 0000 UTC 5 February 1978 through 1200 UTC 7 February 1978. See Fig. 31 for details.

terclockwise loop before it drifted very slowly eastward just south of New England on 7 February. This slow movement prolonged the heavy snowfall from Long Island to eastern New England.

- The heaviest snowfall occurred to the north and northeast of the surface low, as is often the case. Excessive snowfall, however, was also reported to the west and southwest of the center over Maryland and Delaware late on 6 February.

*The 850-mb chart*

- Northwesterly flow and cold air advection preceded the storm in the northeastern United States on 4 February. The 850-mb temperatures generally ranged between $-5°C$ and $-15°C$ over the northeastern United States.
- A weak 850-mb low dropped southeastward across the Great Lakes on 5 February and deepened very gradually. A new 850 mb low center formed just off the Virginia coast by 1200 UTC 6 February. The redevelopment of the 850-mb low was first evident at 0000 UTC 6 February when the largest height falls were located in the Carolinas, well south of the primary center, which was positioned near Buffalo, New York. The secondary low deepened rapidly $[-150 \text{ m } (12 \text{ h})^{-1}]$ as it moved northward close to the New Jersey coast by 0000 UTC 7 February, then drifted to the east without any further deepening.
- A very pronounced pattern of cold air advection was evident to the west of the 850-mb low center prior to and during the development of the secondary low. The $-20°C$ isotherm was driven from Manitoba to North Carolina on 6–7 February. Warm air advection did not develop until 0000 UTC 6 February, over eastern North Carolina and Virginia, as the secondary cyclone was beginning to form over the Atlantic Ocean. An S-shaped isotherm pattern evolved along the East Coast by 1200 UTC 6 February, reflecting the strong cold and warm air advections surrounding the 850-mb low during secondary cyclogenesis.
- A strong easterly LLJ with velocities exceeding 30 m s$^{-1}$ developed to the north of the cyclone center in conjunction with the rapid deepening between 1200 UTC 6 February and 0000 UTC 7 February. The development of this jet coincided with the rapid northward and westward expansion of snowfall over the Middle Atlantic and New England states. The moisture transports and ascending motions associated with the LLJ were apparently important elements in the development of heavy snowfall over the northeastern United States.

*500-mb geopotential height analyses*

- The precyclogenetic environment at 500 mb included a relatively stationary ridge over Greenland, a closed cyclonic circulation over southeastern Canada, a trough across south-central Canada and the Great Lakes, and a ridge over western North America on 5 February. The vortex over southeastern Canada was associated with a confluent geopotential height pattern over New England that was located above the 850-mb cold advection area and sea-level high-pressure ridge.
- The East Coast cyclone developed while the flow regime across North America

# 850 MB

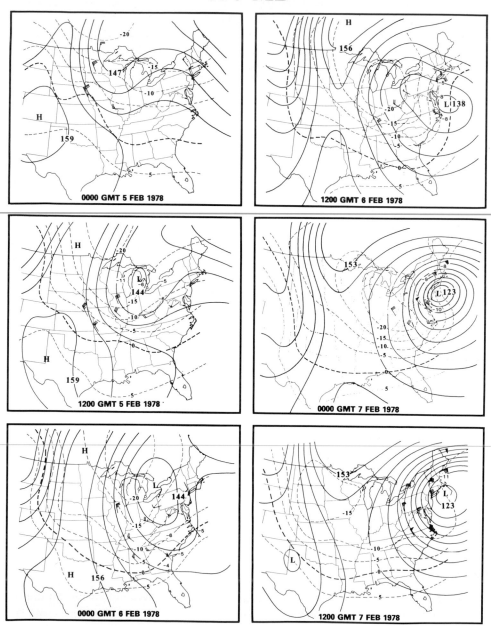

FIG. 97.    Twelve-hourly 850-mb analyses for 0000 UTC 5 February 1978 through 1200 UTC 7 February 1978. See Fig. 32 for details.

underwent a remarkable transformation. The 500-mb pattern was characterized by nearly zonal flow prior to 5 February, by strongly meridional flow on
5–6 February, and by large, symmetrical vortices on 7 February. The upper-
level ridge over the West amplified on 5–6 February and evolved into a closed
anticyclonic circulation in central Canada by 7 February. Concurrently, the
trough in south-central Canada amplified and dropped southeastward, evolving
into a large cyclonic circulation over the eastern United States as the surface
low intensified off the Middle Atlantic coast. This is an excellent example of
a deepening or "digging" trough system (see section 5.1.3).

- The growing cyclonic circulation was accompanied by an amplifying cyclonic
  vorticity maximum that propagated southeastward from the Great Lakes region on 5 February to the Virginia coast on 6 February. This vorticity maximum appeared to be associated with the cyclonic curvature and increasing
  cyclonic wind shears associated with a jet streak propagating southeastward
  over the central United States.
- A diffluent geopotential height pattern developed downstream of the trough
  between 1200 UTC 5 February to 1200 UTC 6 February. The trough axis
  rotated dramatically between 0000 UTC 6 February and 0000 UTC 7 February, the period during which the rapid sea-level development took place.
- Very pronounced amplitude changes also occurred during the cyclogenetic
  period. The amplitude of the trough and upstream ridge increased greatly
  from 0000 UTC 5 February through 0000 UTC 7 February. The latitudinal
  separation of the 5520-m contour at the ridge and trough axes grew from 14°
  to greater than 30° latitude (1500 to 3000 km). The amplitude of the trough
  and downstream ridge increased most rapidly between 1200 UTC 5 February
  and 0000 UTC 7 February, during the period of rapid cyclogenesis.
- The half-wavelength between the trough and the *upstream* ridge increased
  prior to East Coast cyclogenesis on 6 February. The half-wavelength between
  the trough and the *downstream* ridge decreased during 6–7 February while
  cyclogenesis was underway off the coast.
- The intense cyclonic and anticyclonic upper-level vortices established a blocking pattern over the eastern half of North America. The large cyclonic vortex
  drifted from the New England coast into eastern Canada, where it meandered
  for several weeks following the storm. This resulted in persistent cold and
  dry weather over the northeastern United States for the remainder of February 1978.

*Upper-level wind analyses*

- A subtropical jet of 50 m s$^{-1}$ was observed at 200 mb off the Southeast coast
  at 1200 UTC 5 February and 0000 UTC 6 February prior to major cyclogenesis. It appeared to become incorporated into the amplifying upper-level
  trough nearing the East Coast by 0000 UTC 6 February (see the infrared
  satellite sequence).
- A polar jet streak with winds of 60–70 m s$^{-1}$ was positioned upwind of the
  trough axis prior to cyclogenesis. The strongest winds remained just west of
  the diffluent portion of the trough as the system moved toward the East Coast

# 500 MB HEIGHTS AND UPPER-LEVEL WINDS

FIG. 98. Twelve-hourly analyses of 500-mb geopotential heights and upper-level wind fields for 0000 UTC 5 February 1978 through 1200 UTC 7 February 1978. See Fig. 33 for details.

on 5 and 6 February. The cyclone development commenced beneath the diffluence region downwind of the trough axis and within the exit region of the polar jet, where cyclonic vorticity advection and upper-level divergence were maximized.

- At 1200 UTC 6 February, a separate jet formed east of the trough axis and was directed toward the north along the coastline, as heavy snowfall developed.

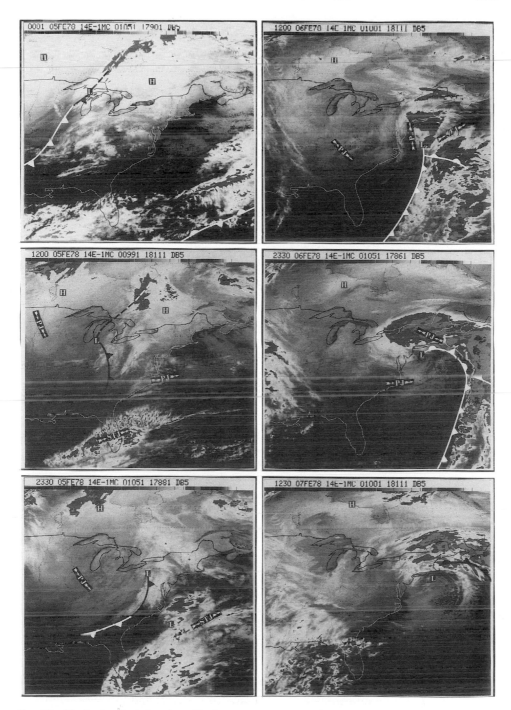

FIG. 99.　Twelve-hourly infrared satellite images for 0000 UTC 5 February 1978 through 1230 UTC 7 February 1978. Surface fronts and low- and high-pressure centers are shown, along with the locations of upper-level polar (PJ) and subtropical (STJ) jet streaks.

The cyclogenesis occurred within the entrance region of this developing jet streak.

*Infrared satellite imagery sequence*

- A diffuse cloud area moved from the Great Lakes toward the eastern United States within the diffluent exit region of the trough/polar jet system prior to 1200 UTC 5 February.
- The 2330 UTC 5 February satellite image depicted the first stage of major cyclogenesis. The cloud mass over the eastern United States was more distinct, with a region of cold cloud tops forming off the North Carolina coast. This developing cloud mass was located east of the upper-level trough axis and was associated with the initial cyclogenesis. An expanding region of cold cloud tops off Florida appeared to be associated with the subtropical jet propagating out over the Atlantic Ocean.
- By 1200 UTC 6 February, the cloud feature off Florida had moved north-eastward over the Atlantic Ocean. Its northern edge displayed the distinctly anticyclonic curvature characteristic of the warm conveyor belt described in section 5.5. The cloud mass originally located off North Carolina expanded northward and then westward over the northeastern states by 0000 UTC 7 February, coinciding with the development of strong easterly flow at lower and upper levels as the 500-mb vortex developed along the Virginia coast. This cloud feature resembles the cold conveyor belt described in section 5.5 and became the comma head within a cloud mass that developed the comma signature by 2330 UTC 6 February. The heaviest snow fell over the Middle Atlantic states and southern New England on 6 and 7 February, along the southern edge of the coldest tops in the comma head.
- As the cyclone occluded off the New England coast, the cloud mass spiraled around the low center by 1230 UTC 7 February. The coldest cloud tops now had become disorganized and were located well to the north of the low center. A pronounced eye-like cloud-free region was observed in visible images on the morning of 7 February (see section 5.5 and Fig. 27), nearly collocated with the positions of the surface low and the 500-mb vortex.

## 7.18    18–20 FEBRUARY 1979

*General remarks*

- This heavy snowstorm hit the Middle Atlantic states with record snow amounts on the "Presidents' Day" holiday. Snowfall rates approached 12 cm h$^{-1}$ on the morning of 19 February (Foster and Leffler 1979). The cyclone has been the subject of numerous studies (Bosart 1981; Bosart and Lin 1984; Uccellini et al. 1984, 1985, 1987; Atlas 1987) since the rapidly developing storm and heavy snowfall rates observed on 19 February were poorly forecast by oper-ational numerical prediction models. The snowstorm occurred at the end of an unusually cold period that culminated in a massive cold air outbreak on 17–18 February and the development of the "Presidents' Day Storm."

- Regions with snow accumulations exceeding 25 cm: portions of West Virginia and Virginia (especially northern Virginia), Maryland, Delaware, southern Pennsylvania, most of New Jersey, and New York City.
- Regions with snow accumulations exceeding 50 cm: parts of Maryland, Delaware, and New Jersey.
- Urban center snowfall amounts:

| | |
|---|---|
| Washington, D.C.–National Airport | 18.7″ (47 cm) |
| Baltimore, MD | 20.0″ (51 cm) |
| Philadelphia, PA | 14.3″ (36 cm) |
| New York, NY–Central Park | 12.7″ (32 cm) |

- Other selected snowfall amounts:

| | |
|---|---|
| Dover, DE | 25.0″ (64 cm) |
| Atlantic City, NJ | 17.1″ (43 cm) |
| Newark, NJ | 16.6″ (42 cm) |
| Wilmington, DE | 16.5″ (42 cm) |
| Harrisburg, PA | 14.2″ (36 cm) |
| Richmond, VA | 10.9″ (28 cm) |

*The surface chart*

- The cyclone was preceded by a massive anticyclone (1050 mb) that established many record low temperatures across the eastern third of the United States.
- Cold air damming occurred to the lee of the Appalachian Mountains on 18 February, as indicated by the pronounced inverted sea-level pressure ridge over the eastern United States at 1200 UTC 18 February and 0000 UTC 19 February. When snow began falling across the Middle Atlantic states on 18 February, surface temperatures generally ranged between $-15°C$ and $-10°C$.
- Two inverted sea-level pressure troughs formed prior to the development of the cyclone. One trough formed from the Gulf coast to the Ohio Valley late on 17 February. The other appeared off the Southeast coast early on 18 February, in conjunction with the development of a pronounced coastal front. Heavy snow, freezing rain, and ice pellets broke out across the southeastern United States to the west of the developing coastal front/inverted trough.
- A primary low-pressure center formed over Kentucky (labeled "primary" because the inverted trough in which it developed formed *prior* to the coastal trough), but had no associated frontal features. This initial low deepened to only 1024 mb over the Ohio Valley and was associated with light to moderate snows just south of the Great Lakes on 18 February. The low dissipated as rapid secondary cyclogenesis commenced along the East Coast early on 19 February.
- The secondary cyclone formed along the coastal front off the Georgia–Florida coast by 1800 UTC 18 February and propagated northeastward to a position near the Maryland coast by 1200 UTC 19 February. Heavy snow developed across the Middle Atlantic states, with snowfall rates of $5–7.5$ cm h$^{-1}$ common from Washington, D.C., to New York City. The low-pressure center made an

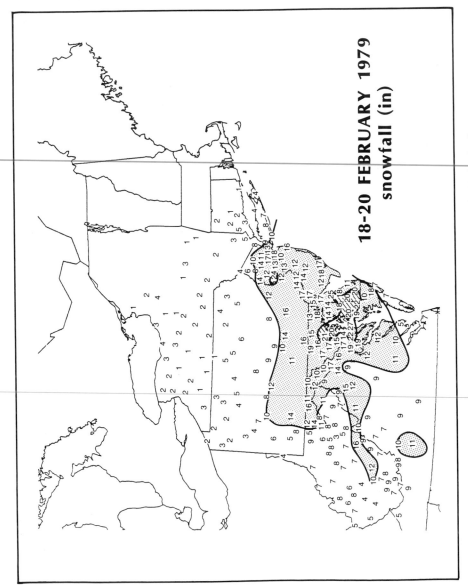

FIG. 100.   Snowfall (in) for 18–20 February 1979. See Fig. 30 for details.

# SURFACE

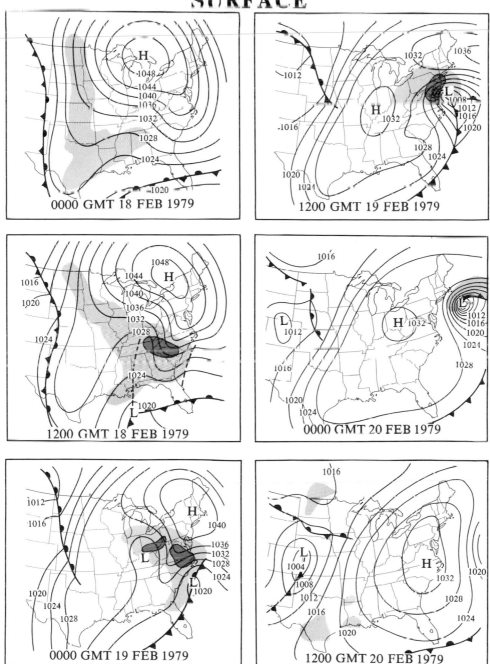

FIG. 101.  Twelve-hourly surface weather analyses for 0000 UTC 18 February 1979 through 1200 UTC 20 February 1979. See Fig. 31 for details.

abrupt turn to the east after 1200 UTC 19 February, sparing New England significant accumulations. The cyclone propagated at an average rate of 15 m s$^{-1}$.

- The storm developed in a regime of very high sea-level pressure and deepened to only 995 mb {based on NMC analyses [Bosart (1981) established a central pressure of 990 mb by 1800 UTC 19 February]} east of the Maryland coast. The heaviest snows fell across the Middle Atlantic states on 19 February, when the deepening rates approached a maximum of −2 to −3 mb h$^{-1}$.

*The 850-mb chart*

- An 850-mb ridge that was associated with the large surface anticyclone dominated the eastern United States prior to East Coast cyclogenesis. Cold air advection and northwesterly flow east of the ridge axis were located beneath a region of confluent geopotential heights at 500 mb. The 850-mb 0°C isotherm was forced as far south as South Carolina, with temperatures below −20°C common across the northeastern United States on 17 February.
- An 850-mb low center formed over the Midwest at 0000 UTC 19 February, well north of the 0°C isotherm and the largest temperature gradient. A separate region of geopotential height-falls developed across the southeastern United States at 1200 UTC 18 February and 0000 UTC 19 February. This height-fall center was located near the 0°C isotherm in a region where heavy precipitation and the coastal inverted trough were developing. The 850-mb low appeared to redevelop over southeastern Virginia by 1200 UTC 19 February, probably in response to an amplification of the geopotential height-fall center as it moved northeastward with the deepening surface low. It may reflect a merger of the Ohio Valley low center with the amplifying coastal height-fall center.
- A southeasterly, ageostrophic LLJ with winds of up to 25 m s$^{-1}$ formed over Georgia and South Carolina by 1200 UTC 18 February. The LLJ appeared at the same time that the coastal front and heavy precipitation developed across the southeastern United States. More detailed case studies linked the development of the LLJ to a combination of interrelated factors. These included a geopotential height-fall center and indirect transverse circulation across the southeastern United States in a region of large 850-mb temperature gradients, sensible and latent heating, and cold air damming near the surface [see Uccellini et al. (1984, 1987)].
- The LLJ shifted further north along the East Coast during the 24 hours ending at 1200 UTC 19 February and was a significant factor in doubling the moisture transports into the region of heavy snowfall on 18 and 19 February (Uccellini et al. 1984). Strong warm air advection was associated with the LLJ between 1200 UTC 18 February and 1200 UTC 19 February, helping to create an S-shaped isotherm pattern along the East Coast.

*500-mb geopotential height analyses*

- A distinct pattern of confluent geopotential heights across the northeastern United States on 17 and 18 February was associated with the movement of

# 850 MB

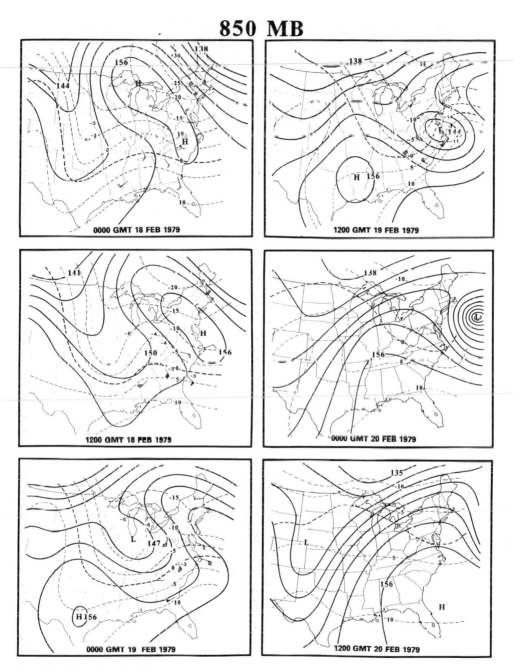

FiG. 102.    Twelve-hourly 850-mb analyses for 0000 UTC 18 February 1979 through 1200 UTC 20 February 1979. See Fig. 32 for details.

a very cold anticyclone from the Great Lakes region to New England. The circulation across the remainder of the United States and southern Canada was dominated by a nearly zonal flow. A low-amplitude trough crossed the central United States from west to east prior to East Coast cyclogenesis.

- Separate troughs, identified by distinct cyclonic vorticity maxima north of Montana and over New Mexico at 1200 UTC 17 February (not shown), appeared to merge into one trough over the Mississippi Valley by 1200 UTC 18 February. Uccellini et al. (1984, 1985) showed that a tropopause fold associated with this trough extruded stratospheric air with high potential vorticity down to the 700-mb level by 0000 UTC 19 February. The trough and its associated vorticity maximum propagated eastward, crossing the East Coast as the surface low deepened rapidly after 1200 UTC 19 February. No closed cyclonic circulation was observed at 500 mb during the evolution of this storm.

- Diffluence became increasingly evident downwind of the trough axis after 0000 UTC 18 February, coinciding with the formation of the inverted troughs and low-pressure centers over the Ohio Valley and Atlantic coast. The orientation of the upper-level trough axis rotated slightly from north–south to northwest–southeast by 1200 UTC 19 February, with a pronounced pattern of diffluent geopotential heights along the East Coast as the secondary cyclone was deepening rapidly off Maryland.

- The amplitude of the trough and downstream ridge increased slightly between 1200 UTC 18 February and 0000 UTC 19 February, but actually appeared to *decrease* slightly in the 12-hour period ending at 1200 UTC 19 February, while the secondary cyclone was intensifying rapidly off the Middle Atlantic coast. Between 0000 UTC 18 February and 1200 UTC 19 February, the distance between the trough axis and the downstream ridge decreased.

- The precipitation area expanded on 18 and 19 February as the half-wavelength between trough and downstream ridge decreased and rapid cyclogenesis commenced off the East Coast.

- Details on the evolution of the upper-level features can be found in Uccellini et al. (1984, 1985). See Bosart (1981) and Bosart and Lin (1984) for surface and quasi-geostrophic analyses.

*Upper-level wind analyses*

- A polar jet attained speeds of up to 80 m s$^{-1}$ near the New England coast by 1200 UTC 18 February. This jet was imbedded within the confluent region to the rear of the upper-level trough moving away from eastern Canada and was located above the large surface anticyclone positioned over the northeastern United States.

- Winds within the subtropical jet streak at 200 mb strengthened near the crest of the ridge approaching the East Coast as heavy snow developed over the southeastern United States early on 18 February, prior to East Coast cyclogenesis. Wind speeds increased from 60 m s$^{-1}$ to greater than 80 m s$^{-1}$ between 1200 UTC 17 February and 1200 UTC 18 February as the flow developed a noticeable cross-contour component near the ridge crest. Increasing upper-

# 500 MB HEIGHTS AND UPPER-LEVEL WINDS

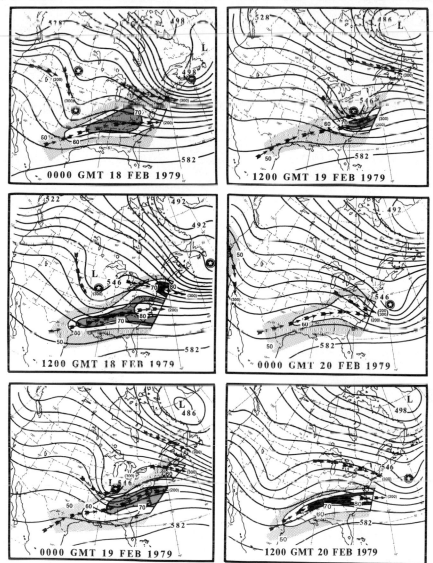

FIG. 103.   Twelve-hourly analyses of 500-mb geopotential heights and upper-level wind fields for 0000 UTC 18 February 1979 through 1200 UTC 20 February 1979. See Fig. 33 for details.

level divergence associated with this subtropical jet contributed to the initial outbreak of precipitation in the southeastern United States (Uccellini et al. 1984).

- A separate 50 m s$^{-1}$ polar jet streak propagated from the northern Plains states to the East Coast between 0000 UTC 18 February and 1200 UTC 19 February. A tropopause fold was diagnosed along the axis of this jet (Uccellini

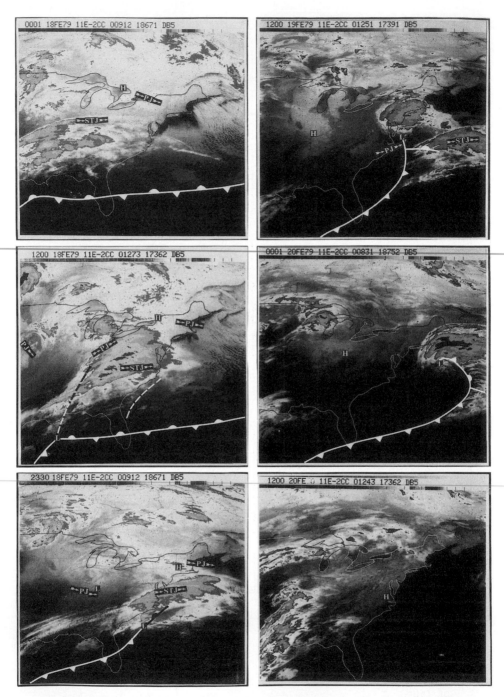

FIG. 104.    Twelve-hourly infrared satellite images for 0000 UTC 18 February 1979 through 1200 UTC 20 February 1979. Surface fronts and low- and high-pressure centers are shown, along with the locations of upper-level polar (PJ) and subtropical (STJ) jet streaks.

et al. 1985). The rapid development of precipitation across the Middle Atlantic states and the explosive surface cyclogenesis occurred within the diffluent exit region of this jet after 0000 UTC 19 February, as the polar jet propagated toward the East Coast.

- See Uccellini et al. (1984, 1985) for a more detailed description of the jet streaks that influenced the development of the "Presidents' Day Storm."

*Infrared satellite imagery sequence*

- The cyclone and heavy snowfall were associated with two separate cloud systems.
- One cold-top cloud mass streamed eastward along the axis of the subtropical jet between 0000 UTC and 2330 UTC 18 February. The heaviest precipitation occurred along the southern edge of this cloud system, with heavy snow and freezing precipitation falling across Kentucky, Tennessee, Georgia, and the Carolinas on 18 February. These clouds moved off the East Coast and diminished in areal coverage as cyclogenesis commenced along the Carolina coast late on 18 February.
- A small comma-shaped cloud area located over Nebraska and Iowa at 1200 UTC 18 February traversed the eastern half of the country in the ensuing 24-hour period. This cloud mass was located in the exit region of the polar jet during this entire period and expanded rapidly in the 6-hour period between 0900 UTC and 1500 UTC 19 February [see Uccellini et al. (1985)]. Much colder cloud-top temperatures blossomed from Virginia to Massachusetts at 1200 UTC 19 February as the surface cyclone deepened rapidly off the Virginia–Maryland coast, with very heavy snows in the Middle Atlantic states. The coldest cloud-top temperatures were found to the north of the surface low. The heaviest snows were located from the coldest cloud-top area near New York City southward to the edge of the cold tops over northern Virginia at 1200 UTC 19 February. The rapid development of this feature resembles a cold conveyor belt (Carlson 1980) and represents the emerging head of a comma-shaped cloud mass which became quite distinct over the Atlantic Ocean by 0001 UTC 20 February.
- A cloud-free "eye" developed over the Atlantic by 1800 UTC 19 February (see Fig. 27) after the storm went through a period of rapid sea-level intensification.

## 7.19   5–7 APRIL 1982

*General remarks*

- This unusual late-season storm produced near-blizzard conditions over much of Pennsylvania, New York, and New England, including New York City and Boston. The rapidly intensifying cyclone spread heavy snow amounts across the Midwest and the Ohio Valley before reaching the northern Middle Atlantic states and New England. The snow and cold temperatures postponed opening day of the major league baseball season in many cities. Thunderstorms with frequent lightning were reported in New York City during the heaviest snowfall. The storm was followed by one of the coldest air masses on record for April. The temperature at Boston, Massachusetts, remained near $-10°C$ during the afternoon of 7 April. Operational numerical simulations were successful in forecasting this storm. [See Kaplan et al. (1982) for a mesoscale simulation of the secondary sea-level development associated with this case.]

- Regions with snow accumulations exceeding 25 cm: portions of northern and eastern Pennsylvania, northern New Jersey, southern New York, central and northern Connecticut and Rhode Island, Massachusetts (except the extreme southeast), southern Vermont and New Hampshire, and Maine.
- Regions with snow accumulations exceeding 50 cm: scattered portions of eastern New York, southern Vermont, northeastern Massachusetts, and southeastern New Hampshire.
- Urban center snowfall amounts:

| | |
|---|---|
| Philadelphia, PA | 3.5" (9 cm) |
| New York, NY–Central Park | 9.6" (24 cm) |
| Boston, MA | 13.3" (34 cm) |

- Other selected snowfall amounts:

| | |
|---|---|
| Grafton, NY | 22.8" (58 cm) |
| Albany, NY | 17.7" (45 cm) |
| Portland, ME | 15.9" (40 cm) |
| Worcester, MA | 15.0" (38 cm) |
| Hartford, CT | 14.1" (36 cm) |
| Concord, NH | 13.9" (35 cm) |
| Newark, NJ | 12.8" (33 cm) |
| Allentown, PA | 11.4" (29 cm) |

*The surface chart*

- Relatively cold air moved into the northeastern United States behind an intense cyclone that propagated across eastern Canada. Unseasonably cold temperatures were associated with an anticyclone (1033 mb) located over central Canada. A sea-level high-pressure ridge extending southeastward from the anticyclone shifted from the Ohio Valley to the northeastern United States on 5 April, providing cold air for the snowstorm which developed the next day. Following the snowstorm, the anticyclone moved southward into the Midwest. Its effects combined with the snow cover to produce record low temperatures from the Plains states to the East Coast on 6–8 April.
- Only weak cold air damming was observed along the East Coast prior to and during cyclogenesis on 5–6 April.
- Coastal frontogenesis was evident late on 5 April and early on 6 April along the Carolina coast, but was not characterized by the large temperature contrasts seen in most of the other cases.
- A primary low-pressure system moved rapidly from the southern Plains through the Ohio Valley to Pennsylvania between 1200 UTC 5 April and 1200 UTC 6 April. The center deepened 10 mb in the 12-hour period ending

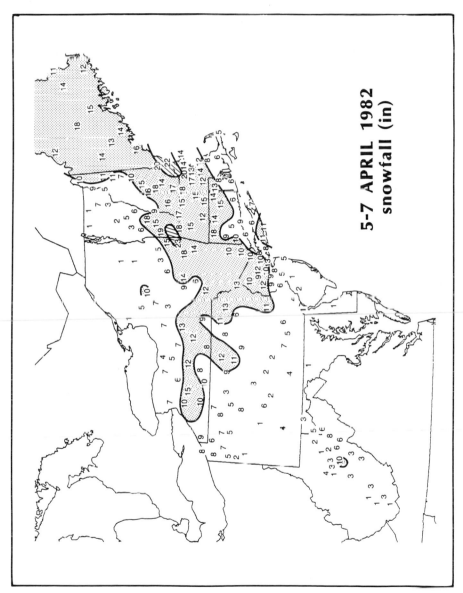

FIG. 105.  Snowfall (in) for 5–7 April 1982. See Fig. 30 for details.

# SURFACE

FIG. 106. Twelve-hourly surface weather analyses for 0000 UTC 5 April 1982 through 1200 UTC 7 April 1982. See Fig. 31 for details.

at 0900 UTC 6 April. The primary low produced moderate to heavy snows and strong winds from Chicago to Detroit and Cleveland as it moved through the Ohio Valley on 5–6 April. Although this case exhibited the most intense primary surface system of any in the sample, the primary low was very quickly absorbed into the circulation of the secondary coastal storm on 6 April.

- A weak low-pressure center was observed on the coastal front near the Georgia coast by 0000 UTC 6 April. Explosive cyclogenesis commenced further to the north, however, along the Virginia coast, between 0000 UTC and 1200 UTC 6 April. The storm deepened at a rate of $-1$ to $-3$ mb h$^{-1}$, with the central sea-level pressure falling from 994 mb at 0900 UTC 6 April to 968 mb by 0000 UTC 7 April. The low center tracked east-northeastward from the Virginia–Maryland coast, passing south of Long Island and New England, then turned northeastward toward Nova Scotia on 7 April. The forward motion of the storm maintained a rate of about 15 m s$^{-1}$ throughout this period.
- Over the northeastern United States, most of the snow fell between 1200 UTC 6 April and 1200 UTC 7 April. In many locations, snowfall rates of 2.5–5 cm h$^{-1}$ occurred in conjunction with north-northeasterly surface winds increasing to 15–20 m s$^{-1}$ with higher gusts.

*The 850-mb chart*

- Prior to East Coast cyclogenesis, low-level northwesterly flow behind the intense cyclone in eastern Canada generated strong cold air advection across the northeastern United States on 4–5 April. The northwesterly flow and cold air advection were located beneath a region of upper-level confluence. The strong cold advection helped to create a narrow, intense temperature gradient over the Middle Atlantic states on 5 April.
- The 850-mb low intensified at an accelerating rate as it moved from the central Plains to near southern New England between 1200 UTC 5 April and 0000 UTC 7 April. It deepened at rates of $-60$ m (12 h)$^{-1}$ as it crossed the upper Mississippi Valley by 0000 UTC 6 April, $-90$ m (12 h)$^{-1}$ as it approached the East Coast by 1200 UTC 6 April, and $-150$ m (12 h)$^{-1}$ as it moved off the coast by 0000 UTC 7 April.
- A secondary 850-mb geopotential height-fall center appeared over the southeastern United States at 0000 UTC 6 April, possibly reflecting the secondary surface low development. This feature was difficult to follow at 1200 UTC 6 April since the entire 850-mb circulation had intensified over the eastern United States.
- This case was marked by very strong temperature gradients during cyclogenesis. Pronounced cold air advection occurred behind the low, especially at 1200 UTC 5 April and 0000 UTC 6 April, with strong warm air advection ahead of it, especially at 1200 UTC 6 April and 0000 UTC 7 April. The advection patterns intensified as the 850-mb low moved toward the East Coast on 6 April, producing an S-shaped isotherm pattern.
- The 0°C isotherm was generally located near the 850-mb low center except by 1200 UTC 7 April, when the low was centered in the colder air as it occluded.

# 850 MB

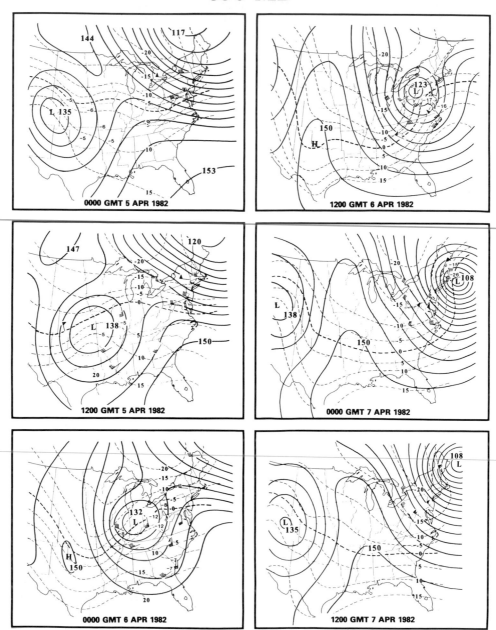

FIG. 107.    Twelve-hourly 850-mb analyses for 0000 UTC 5 April 1982 through 1200 UTC 7 April 1982. See Fig. 32 for details.

- No clearly-defined LLJ formation was observed since high wind speeds were associated with the 850-mb low throughout the study period. The southerly flow to the southeast of the center and the easterly flow to the north of the center both strengthened, however, as the 850-mb low intensified rapidly on 6 April.

*500-mb geopotential height analyses*

- Prior to 6 April, the precyclogenetic environment at 500 mb was characterized by a developing anticyclonic circulation near Greenland and a high-amplitude trough over southeastern Canada. This configuration produced a confluent geopotential height pattern across the Great Lakes and northeastern United States, as the surface anticyclone and cold air became entrenched over the Great Lakes region on 5–6 April.
- The trough associated with the cyclone propagated eastward with the strong westerly flow across the United States from 4 to 7 April. The trough system was small in amplitude initially, but amplified and formed a deepening cyclonic vortex at 500 mb by 0000 UTC 7 April.
- The amplifying trough was associated with an intensifying cyclonic vorticity maximum as it crossed the United States. Increasing vorticity advection accompanied the rapid development of the surface low and the outbreak of heavy snowfall on 6 April.
- Geopotential height gradients increased at the base of the trough between 0000 and 1200 UTC 6 April as both the primary and secondary surface lows intensified. At the same time, diffluence became more evident downwind of the trough, as the trough axis became oriented from northwest to southeast (a negative tilt) by 1200 UTC 6 April. The surface low deepened rapidly and heavy precipitation developed along the coast as the diffluence pattern downwind of the negatively tilted 500-mb trough became better defined and moved across the coast on 6 April.
- The amplitude between the trough and downstream ridge increased between 1200 UTC 5 April and 0000 UTC 7 April, although only a small increase in amplitude occurred during rapid cyclogenesis on 6 April. The half-wavelength between the trough and downstream ridge decreased throughout the 36-hour period ending at 0000 UTC 7 April.

*Upper-level wind analyses*

- A polar jet stretched across the northeastern United States on 5–6 April, within the confluent region upwind of the trough system over southeastern Canada. This confluent jet pattern was located above the cold surface anticyclone over the Great Lakes region. A subtropical jet exited the southeastern United States on 5 April.
- A separate polar jet with maximum wind speeds exceeding 70 m s$^{-1}$ extended from the West Coast into the base of the upper-level trough as the trough amplified over the central United States on 5 April. The increasing cyclonic vorticity maximum was located in the cyclonic exit region of the jet, at the base of the 500-mb trough.

# 500 MB HEIGHTS AND UPPER-LEVEL WINDS

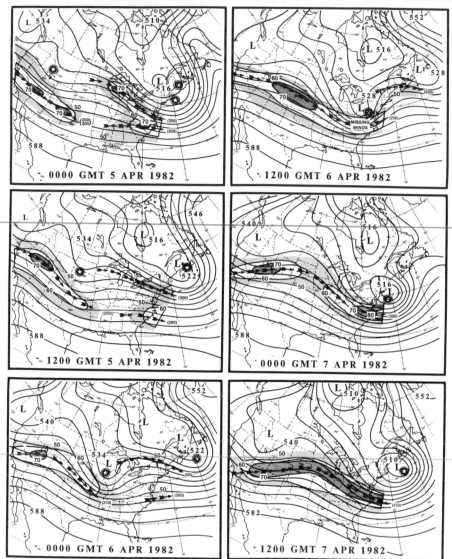

FIG. 108.   Twelve-hourly analyses of 500-mb geopotential heights and upper-level wind fields for 0000 UTC 5 April 1982 through 1200 UTC 7 April 1982. See Fig. 33 for details.

- Missing wind reports over the Carolinas at 1200 UTC 6 April and an 80 m s$^{-1}$ measurement at 0000 UTC 7 April indicate that wind speeds increased near and upwind of the base of the trough immediately before the rapid development of the secondary cyclone on 6 April.
- The storm system developed rapidly within the diffluent exit region of the

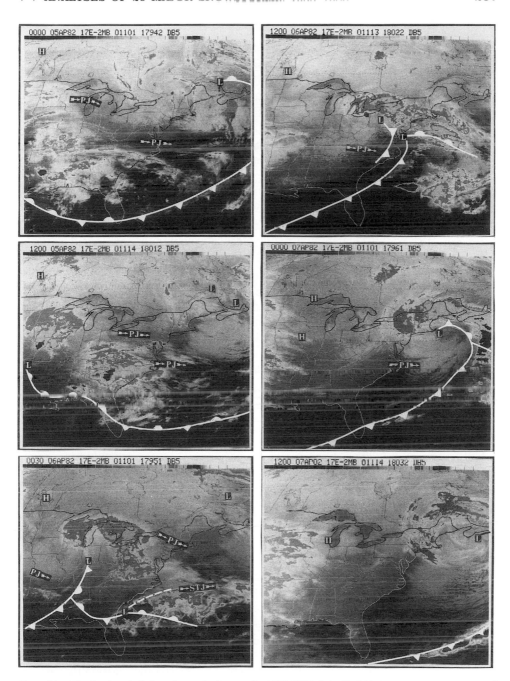

FIG. 109.  Twelve-hourly infrared satellite images for 0000 UTC 5 April 1982 through 1200 UTC 7 April 1982. Surface fronts and low- and high-pressure centers are shown, along with the locations of upper-level polar (PJ) and subtropical (STJ) jet streaks.

polar jet streak/trough system as it approached the East Coast between 0600 and 1200 UTC 6 April, as modeled by Kaplan et al. (1982).

*Infrared satellite imagery sequence*

- A distinct cloud mass with cold cloud tops moved through the southeastern United States between 0000 UTC 5 April and 0000 UTC 6 April. These upper-level clouds were positioned well in advance of the amplifying upper-level trough crossing the United States, and may have been related to the wind maximum passing off the Southeast coast during this period. The cold cloud region was associated with an area of light to moderate rainfall across the southeastern United States. By 0000 UTC 6 April, it was located near the weak surface low along the Southeast coast.

- A separate cloud mass extending from Missouri to Minnesota at 1200 UTC 5 April was associated with the upper-level trough and was located north of the surface low over Oklahoma. This cloud mass remained in the diffluent exit region of the polar jet as it propagated eastward. The comma shape of the cloud mass was evident as early as 1200 UTC 5 April, in association with the development of the primary cyclone. There was no marked redevelopment of the comma head as the secondary cyclone deepened after 1200 UTC 6 April. The cloud mass consolidated into a more distinct comma-shaped cloud feature by 0000 UTC 7 April, however, as the surface low intensified rapidly off the East Coast. The heaviest snows fell along the southern edge of this cloud feature.

- The comma cloud evolved into the spiraling cloud mass which characterized the occlusion stage of the cyclone by 1200 UTC 7 April. An eye was briefly visible late on 6 April as the cyclone was deepening rapidly south of New England (see Fig. 27).

## 7.20  10–12 FEBRUARY 1983

*General remarks*

- This snowstorm was one of many cyclones to affect the eastern United States during a winter that was unusually warm and stormy, but it was one of the few storms accompanied by temperatures cold enough for snowfall. The heaviest snows from this storm were oriented near a line through Washington, D.C., Baltimore, Maryland, Philadelphia, Pennsylvania, New York, New York, and Boston, Massachusetts. The 24-hour snowfalls at Philadelphia, Pennsylvania, Harrisburg, Pennsylvania, Allentown, Pennsylvania, and Hartford, Connecticut, were the greatest on record. For many other cities, this was one of the heaviest snowstorms on record. Accumulations reached 75 cm in parts of northern Virginia, western Maryland, and the panhandle of West Virginia. Given the widespread severe effects of this storm in the heavily populated Northeast urban corridor, Sanders and Bosart (1985a,b) named this the "Megalopolitan Snowstorm." Thunderstorms were observed at many locations from Washington, D.C., to New York City during the afternoon and evening of 11 February. Winds were not as crippling as in some other storms, but wind speeds were still high enough to create blizzard or near-blizzard conditions from eastern Pennsylvania and northern Delaware to Massachusetts.

- Regions with snow accumulations exceeding 25 cm: West Virginia (except the extreme southwest and northwest), Virginia (except the extreme southwest and Tidewater), Maryland (except the extreme southeast), Delaware, the southeastern half of Pennsylvania, New Jersey, southeastern New York, Connecticut, Rhode Island, Massachusetts, and extreme southern New Hampshire.
- Regions with snow accumulations exceeding 50 cm: northern Virginia, northeastern West Virginia, central and northern Maryland, southeastern Pennsylvania, central and northern New Jersey, southeastern New York, central Connecticut, and parts of Massachusetts.
- Urban center snowfall amounts:

  | | |
  |---|---|
  | Washington, D.C.–Dulles Airport | 22.8″ (58 cm) |
  | Baltimore, MD | 22.8″ (58 cm) |
  | Philadelphia, PA | 21.3″ (54 cm) |
  | New York, NY–La Guardia Airport | 22.0″ (56 cm) |
  | Boston, MA | 13.5″ (34 cm) |

- Other selected snowfall amounts:

  | | |
  |---|---|
  | Woodstock, VA | 32.0″ (81 cm) |
  | Allentown, PA | 25.2″ (64 cm) |
  | Harrisburg, PA | 25.0″ (64 cm) |
  | Hartford, CT | 21.0″ (53 cm) |
  | Roanoke, VA | 18.6″ (47 cm) |
  | Richmond, VA | 17.7″ (45 cm) |

*The surface chart*

- A 1040-mb anticyclone drifted very slowly from eastern Ontario into southwestern Quebec while the cyclone was propagating across the southeastern United States on 10–11 February. This anticyclone brought one of the few colder-than-normal air masses to affect the United States during the entire winter.
- The distinct high-pressure ridge from New York to North Carolina indicated cold air damming on 10–11 February. Coastal frontogenesis was observed along the coast of the Carolinas prior to 1200 UTC 10 February.
- The surface low developed along the Gulf coast on 9 February and moved eastward to southern Georgia by 0000 UTC 11 February. It then propagated northeastward along the coastal front near the Southeast coast during 11 February, before turning more to the east-northeast after passing the Virginia coast at 0000 UTC 12 February. The low appeared to jump or redevelop from southern Georgia to near the South Carolina coast early on 11 February. Heavy snow developed in the mountains of North Carolina and in southern and western Virginia after 0000 UTC 11 February and spread through the

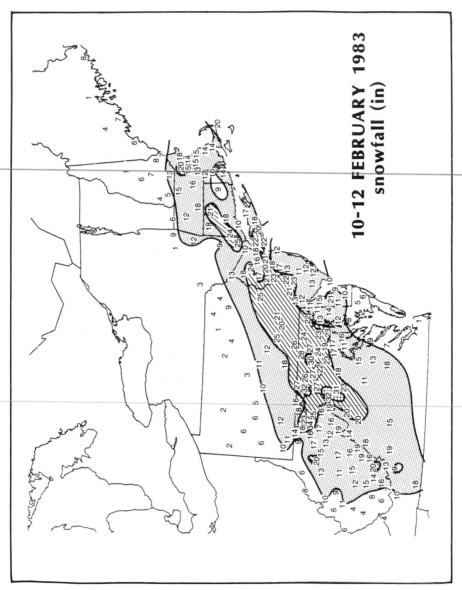

Fig. 110.   Snowfall (in) for 10–12 February 1983. See Fig. 30 for details.

urban centers of the Northeast coastal region during 11 February. The forward speed of the cyclone varied between 10 and 20 m s$^{-1}$.

- The storm deepened slowly and erratically as it propagated from Georgia to near southeastern Virginia on 10–11 February with the central sea-level pressure remaining above 995 mb. The cyclone deepened rapidly only when it had moved east of New England after 1200 UTC 12 February. Pressures never fell below 1012 mb in the major cities of the Northeast, with oscillating pressure tendencies observed from Washington, D.C., to Boston, Massachusetts, during the course of the storm. Evidence presented by Bosart and Sanders (1986) indicates that the erratic pressure tendencies and an outbreak of thunderstorms from Washington, D.C., to New York City were associated with a gravity wave. Snowfall rates of 5–12 cm h$^{-1}$ were reported during the convective outbreak.
- The distance between the low- and high-pressure centers along the Atlantic seaboard decreased as the storm moved up the coast on 11 February. This increased the sea-level pressure gradient north of the low center and strengthened the easterly flow from the Atlantic toward the region of maximum precipitation.

*The 850-mb chart*

- Strong northwesterly flow and cold air advection at 850 mb covered the northeastern United States through 10 February, beneath a region of confluent upper-level geopotential heights. A large 850-mb temperature gradient was established along the East Coast, with temperatures ranging from 0°C over central North Carolina to −20°C over Maine.
- Cold air remained over New England while south to southeasterly winds and warm air advection increased near the Southeast and Middle Atlantic coasts, where 850-mb heights were falling significantly. As a result, temperature gradients along the Southeast and Middle Atlantic coasts increased sharply at 0000 and 1200 UTC 11 February. At these times, moderate to heavy precipitation developed to the west of the coastal front, over the regions experiencing the largest warm air advection at 850 mb.
- The increasing temperature gradients were associated with the development of an S-shaped isotherm pattern along the East Coast between 1200 UTC 11 February and 0000 UTC 12 February. The 850-mb low center was located on the warm side of the 0°C isotherm through 0000 UTC 11 February, but became collocated with the 0°C isotherm as it approached the Atlantic coast at 1200 UTC 11 February.
- The 850-mb low alternately deepened and filled slightly between 0000 UTC 10 February and 1200 UTC 11 February. It then deepened slowly after 1200 UTC 11 February as it moved off the East Coast.
- Although this storm produced some of the deepest snow amounts of any of the cases examined, the LLJs associated with it were only of moderate intensity. Easterly winds to the north of the 850-mb low center were measured at only 20 m s$^{-1}$, although several key reports were missing. Nevertheless, the increasing southeasterly to easterly winds along the coast probably enhanced

# SURFACE

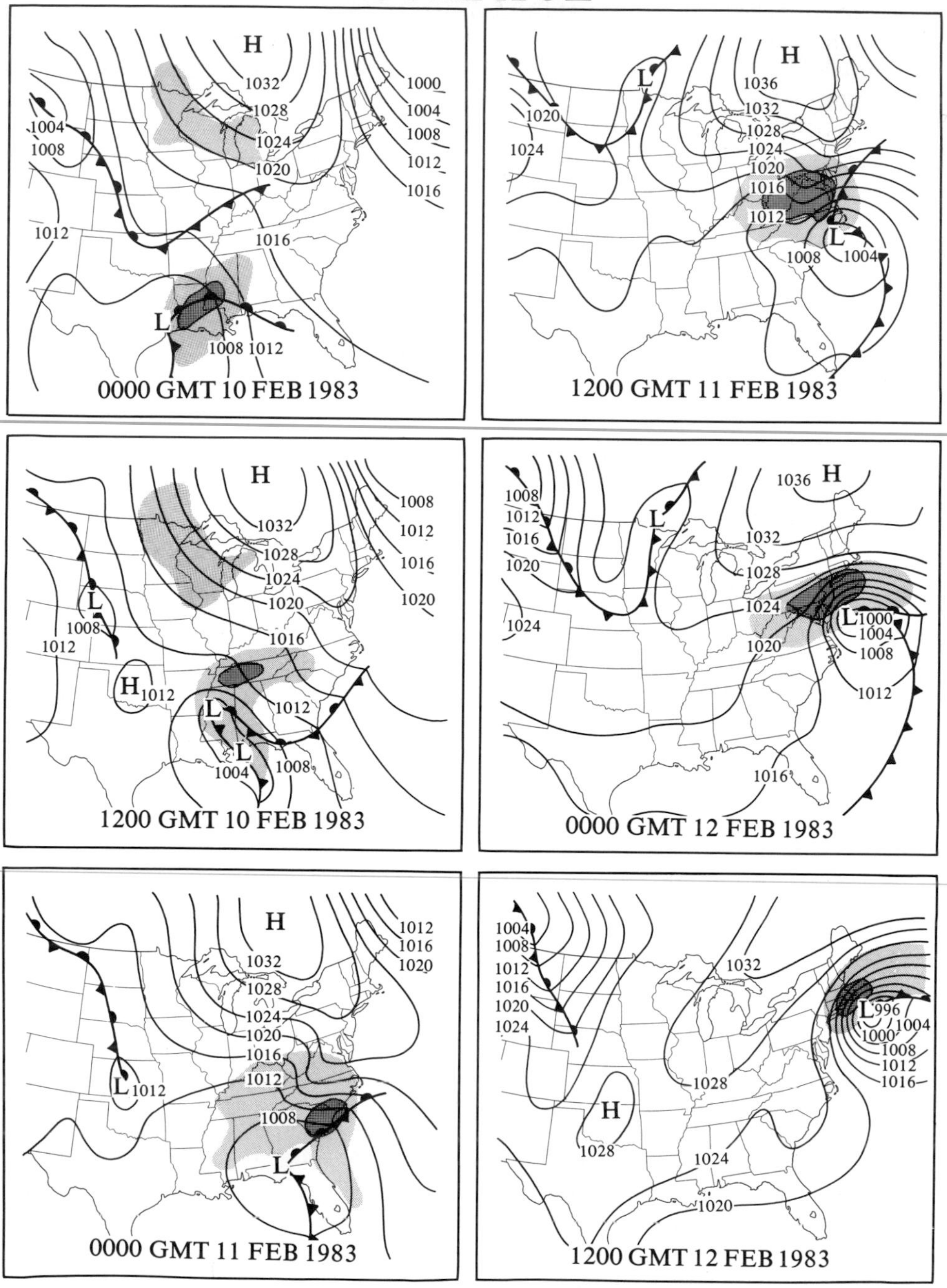

Fig. 111.   Twelve-hourly surface weather analyses for 0000 UTC 10 February 1983 through 1200 UTC 12 February 1983. See Fig. 31 for details.

# 850 MB

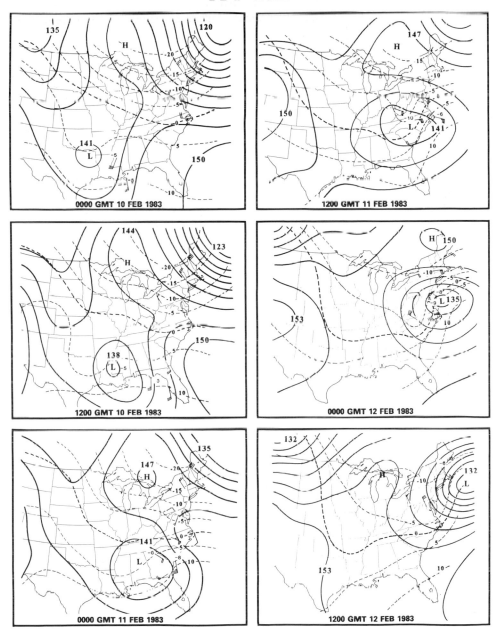

FIG. 112.    Twelve-hourly 850-mb analyses for 0000 UTC 10 February 1983 through 1200 UTC 12 February 1983. See Fig. 32 for details.

the moisture transport into the region of heavy snowfall on 11–12 February. This effect was aided by the slow movement of the 850-mb low, averaging less than 15 m s$^{-1}$, which maintained a strong moisture flow into the Middle Atlantic region for a protracted period.

*500-mb geopotential height analyses*

- Prior to cyclogenesis along the East Coast, an intense, nearly stationary 500-mb anticyclone east of Greenland and a slow-moving trough across southeastern Canada dominated the circulation pattern. The trough in Canada combined with the general westerly flow across the northern United States to produce a pronounced confluent geopotential height pattern over the Great Lakes and northeastern United States on 10–11 February. The surface anticyclone and low-level cold air advection were located beneath this confluence zone.
- A basically zonal flow pattern with several weak troughs across the United States amplified on 11 February. A ridge evolved over the west with two separate well-defined troughs east of the Rocky Mountains. One trough was associated with the East Coast storm while the other produced a new low center in the Gulf of Mexico on 12 February.
- The trough that accompanied the East Coast storm was associated with the merger of two distinct cyclonic vorticity maxima, one propagating across the Gulf states and the other moving east-southeastward across the central Plains on 10 February.
- Between 0000 UTC 10 February and 0000 UTC 11 February, a pronounced pattern of diffluent geopotential heights became established immediately downwind of the trough axis over the southeastern United States. The surface low and heavy precipitation developed within this diffluence region in conjunction with cyclonic vorticity advection along the East Coast.
- The amplitude of the trough and downstream ridge increased only slightly through 1200 UTC 11 February. A closed cyclonic circulation appeared briefly over the Middle Atlantic states at 0000 UTC 12 February.
- The half-wavelength between the trough and the downstream ridge decreased, especially between 1200 UTC 11 February and 0000 UTC 12 February, when heavy snow was falling across the Middle Atlantic states and southern New England.
- More detailed diagnostic analyses of this case examining the influence of gravity wave phenomena, the possible roles of frontogenetical forcing and symmetric instability, and a detailed Doppler radar analysis can be found in Emanuel (1985), Bosart and Sanders (1986), and Sanders and Bosart (1985a,b). A description of the upper-level jet streaks and their associated vertical transverse circulations can be found in Uccellini and Kocin (1987).

*Upper-level wind analyses*

- A 50–60 m s$^{-1}$ polar jet located within the confluent flow over the northeastern United States coincided with the southeastward extension of the surface anticyclone over New England on 10–11 February. Uccellini and Kocin (1987)

# 500-MB HEIGHTS AND UPPER-LEVEL WINDS

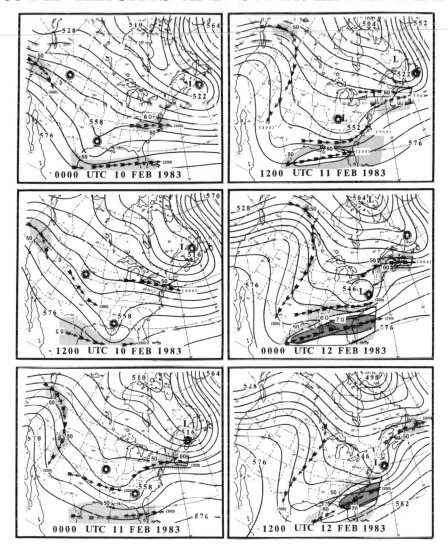

FIG. 113.  Twelve-hourly analyses of 500-mb geopotential heights and upper-level wind fields for 0000 UTC 10 February 1983 through 1200 UTC 12 February 1983. See Fig. 33 for details.

show that the direct transverse circulation in the confluent entrance region of this jet contributed to the low-level cold air advection pattern in the northeastern United States.

- Unlike many of the other cases, the jets propagating into the western portion of the trough that was associated with the East Coast storm were relatively weak (less than 50 m s$^{-1}$), especially before the merger of the two short-wave features at 0000 UTC 11 February.

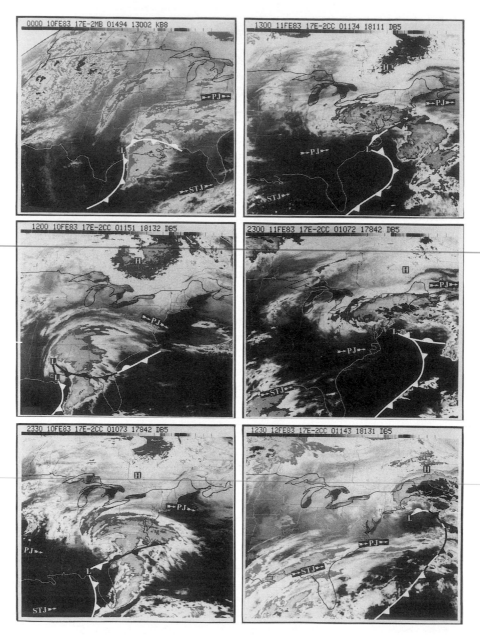

FIG. 114.   Twelve-hourly infrared satellite images for 0000 UTC 10 February 1983 through 1230 UTC 12 February 1983. Surface fronts and low- and high-pressure centers are shown, along with the locations of upper-level polar (PJ) and subtropical (STJ) jet streaks.

- Wind speeds at 200–250 mb within the jet streak near the base of the trough increased over the southeastern United States after 1200 UTC 11 February, as the cyclone developed along the Atlantic coast.

- As shown by Uccellini and Kocin (1987), an indirect circulation in the exit region of the polar jet streak over Georgia at 1200 UTC 11 February accentuated the low-level warm air advection and moisture transport toward the region of heavy snow over the Middle Atlantic states. The rising branches of the transverse circulations in the diffluent exit region of the jet streak in the Southeast and in the confluent entrance region of the jet streak over New England combined to produce a sloped pattern of ascent collocated with the region of heavy snowfall.
- A jet streak with pronounced cross-contour flow propagated from the ridge crest in southwestern Canada toward the base of the second major trough over the southern United States by 0000 UTC 12 February. These features were associated with the development of a new cyclone over the Gulf of Mexico late on 12 February.

*Infrared satellite imagery sequence*

- The shape of the cloud mass associated with the storm changed character during the life of the cyclone. At 0000 UTC 10 February, the cloud mass was marked by a distinct western edge and a small, organized center over Louisiana, similar to Carlson's (1980) schematic for the initial stages of cyclone development. A distinct comma-shaped pattern emerged at 1200 and 2330 UTC 10 February as the surface low moved slowly eastward along the Gulf coast. The northern edge of the cloud shield was generally aligned with the axis of the confluent polar jet over the northeastern United States. At 2330 UTC 10 February, the comma-shaped cloud over the eastern United States was located downwind of the 500-mb cyclonic vorticity maximum over Georgia. The larger cloud mass extended westward to a small comma-shaped cloud pattern over the Kansas–Missouri border which was associated with a separate vorticity maximum located there.
- As the two cyclonic vorticity maxima merged into one over Kentucky by 1200 UTC 11 February, the cloud mass changed character again, stretching into a narrower east–west band of cold cloud tops in the 1300 UTC 11 February image. This band moved slowly northward into the northeastern United States by 2300 UTC 11 February, then drifted eastward off the New England coast by 1230 UTC 12 February, as heavy snow was ending across southeastern New England. The heaviest snow occurred along the southern edge of the high cloud tops associated with this band.
- The very cold cloud tops over New Jersey, southeastern New York, and New England at 2300 UTC 11 February reflect the imbedded convection that was associated with the gravity wave and produced very heavy snowfall.
- The position of the surface cold front and low center corresponded well with the back edge of the enhanced cloud mass at 0000 and 1200 UTC 10 February. At later times, the surface low center was displaced south and west of the region of high cloud tops, as a wedge of dry air moved over the storm center from the southwest.

# CHAPTER 8

# Analyses of Snowstorms during 1987

The year 1987 will be remembered for its many heavy snowstorms over the Middle Atlantic states and southern New England. During the first few weeks of the year, severe winter weather was confined to the interior sections of New York, Pennsylvania, and New England, where near-record amounts of snow had fallen since mid-November 1986. Little snow had occurred along the coast, but the center of activity for major snowstorms soon shifted to the Middle Atlantic region, bringing the area from Virginia through southeastern New England the greatest amounts from late January through February. Storms on 22–23 January, 25–26 January, 9–10 February, 16–18 February, and 22–23 February were the most noteworthy events during this period. The storms of 22–23 January, 25–26 January, and 22–23 February produced major snow accumulations across some of the densely populated regions of the Middle Atlantic states, so map analyses for these cases will be presented in this section. Although the areal extent of heavy snowfall for the two latter cases was somewhat less than the storms sampled earlier in this chapter, their severe impact on major metropolitan areas of the northeastern United States warrants their inclusion.

A rapidly developing cyclone on 9–10 February produced blizzard or near-blizzard conditions, with 25-cm accumulations along the immediate coast from eastern Maryland to southern New Jersey, and 25 cm to more than 50 cm across southeastern Massachusetts. The heaviest snowfall with this system, however, remained outside the populous urban centers.

Late in the year, two exceptional autumn snowstorms left record-breaking snow accumulations across varied sections of the northeastern United States. The first storm, on 3–4 October, was one of the earliest on record across interior eastern New York and western New England. Elevated terrain received accumulations of 50 cm or greater. Then, on 10–12 November, the "Veteran's Day Snowstorm" produced a widespread area of record-setting early-season snowfall from Virginia to Massachusetts. Lightning and thunder accompanied snow accumulations reaching 25–40 cm in the Washington, D.C., area, while 20–30 cm fell in and around Boston, Massachusetts. Another storm in the last week of December left a 25- to 50-cm mantle across southeastern Massachusetts, their third crippling snowfall of the year.

## 8.1  21–23 JANUARY 1987

*General remarks*

- This storm yielded heavy snowfall from Alabama and Georgia to New York and portions of New England. New York City, Philadelphia, Pennsylvania,

Baltimore, Maryland, and Washington, D.C., experienced their largest snowfall in nearly 4 years. Snow accumulations were underpredicted by many fore-casters who anticipated a changeover from snow to rain. In New York City, the changeover did occur, but not until 20–30 cm of snow had already fallen. Some lightning and thunder were reported from Virginia to New York on 22 January.

- Regions with snow accumulations exceeding 25 cm: Virginia (except the southeast), the southeastern half of West Virginia, Maryland (except the eastern shore), northern Delaware, central and eastern Pennsylvania, western and northern New Jersey, central and eastern New York, northwestern Connecticut, western Massachusetts, much of New Hampshire and Vermont, and scattered areas of Maine.

- Regions with snow accumulations exceeding 50 cm: scattered sections of Virginia, Pennsylvania, and New York.

- Urban center snowfall amounts:

  | | |
  |---|---|
  | Washington, D.C.–Dulles Airport | 11.1″ (28 cm) |
  | Baltimore, MD | 12.3″ (31 cm) |
  | Philadelphia, PA | 8.8″ (22 cm) |
  | New York, NY–La Guardia Airport | 11.3″ (29 cm) |
  | Boston, MA | 5.3″ (13 cm) |

- Other selected snowfall amounts:

  | | |
  |---|---|
  | Albany, NY | 16.6″ (42 cm) |
  | Williamsport, PA | 15.8″ (40 cm) |
  | Patuxent River, MD | 13.0″ (33 cm) |
  | Lynchburg, VA | 12.3″ (31 cm) |
  | Wilmington, DE | 12.1″ (31 cm) |
  | Harrisburg, PA | 11.4″ (29 cm) |
  | Allentown, PA | 11.1″ (28 cm) |
  | Newark, NJ | 11.0″ (28 cm) |
  | Hartford, CT | 10.2″ (26 cm) |

*The surface chart*

- The cold air for this snowstorm was associated with a relatively weak anti-cyclone (1022 mb) that propagated from the southern Plains northeastward across the Middle Atlantic states and New England to eastern Canada. This differs markedly from the large, intense anticyclones (>1035 mb) moving across southern Canada which characterized most of the other cases (see Chapter 4).

- An inverted sea-level high-pressure ridge extending from New Jersey to North Carolina indicated cold air damming at 1200 UTC 22 January. The appearance of this high-pressure ridge coincided with the development of a pronounced coastal front parallel to the southeastern United States coast, along which the surface cyclone advanced.

- As the cyclone developed in the western Gulf of Mexico, it moved very slowly, averaging only 10 m s$^{-1}$ between 1200 UTC 21 January and 0000 UTC 22

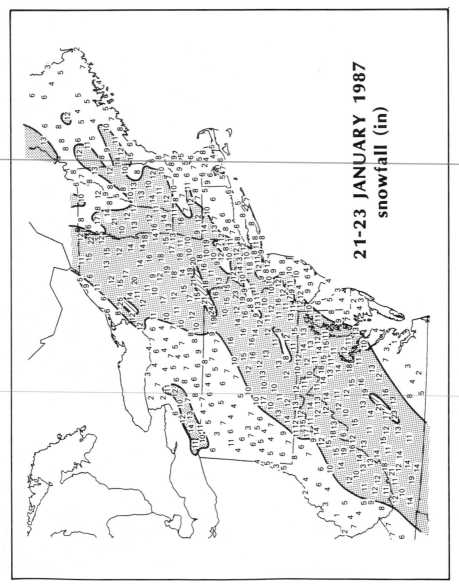

FIG. 115.   Snowfall (in) for 21–23 January 1987. See Fig. 30 for details.

# SURFACE

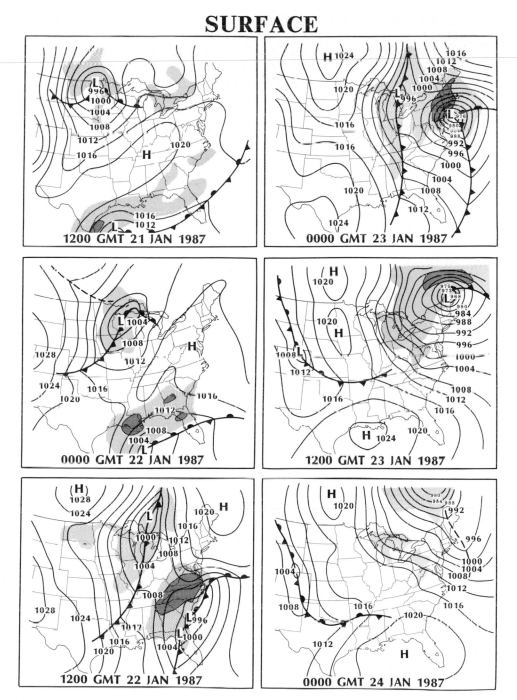

FIG. 116.   Twelve-hourly surface weather analyses for 1200 UTC 21 January 1987 through 0000 UTC 24 January 1987. See Fig. 31 for details.

January. The low center then accelerated northeastward as it neared northern Florida, averaging 18–20 m s$^{-1}$ in the 12-hour period ending at 1200 UTC 22 January. As the cyclone reached northern Florida, a new center formed to its northeast along the coastal front off Georgia and South Carolina. Moderate to heavy snow spread from Georgia to Virginia and Maryland as the new center developed. The initial low was absorbed into the new circulation as it began deepening rapidly.

- The low center moved northeastward along the coastal front through eastern North Carolina on 22 January, reaching a position just off the New Jersey coast by 0000 UTC 23 January. The center advanced at a rate of 18–20 m s$^{-1}$ over the 12-hour period ending at 0000 UTC 23 January. The track of the low remained close to the coastline, causing a changeover from snow to rain along the immediate coast of the Middle Atlantic states. The cyclone then crossed central Long Island and central and eastern New England, reaching southern Maine by 1200 UTC 23 January. The forward speed of the storm slowed as it crossed New England, averaging only 12 m s$^{-1}$. The onshore path of the center brought a changeover from snow to rain across eastern New England. Accumulations of greater than 10 cm generally preceded the changeover, except in extreme southeastern Massachusetts, where little snow fell.

- A 300-km-wide band of 25-cm snowfall was oriented nearly parallel to the path of the surface low. The easternmost edge of the 25-cm accumulations was generally located between 30 and 160 km west of the low track.

- The cyclone deepened slowly as it crossed the Gulf of Mexico, with the central pressure falling from 1007 mb at 1200 UTC 21 January to 1003 mb by 0000 UTC 22 January, and to 998 mb by 1200 UTC 22 January. As the low redeveloped and accelerated along the coastal front and moved to a position east of New Jersey, it deepened 20 mb in only 12 hours, reaching 975 mb by 0000 UTC 23 January. The storm continued to intensify as it crossed Long Island and New England, attaining a minimum pressure of 966 mb over Maine. The rapid development phase was accompanied by snowfall rates of 2.5–7.5 cm h$^{-1}$ from North Carolina to New England.

- A dissipating cold front ushered in the coldest air of the season in the wake of the storm, resulting in persistent winter weather conditions across the Middle Atlantic states and New England.

*The 850-mb chart*

- A wedge of cold 850-mb temperatures ($-10°C$ to $-15°C$) advanced from the Great Lakes into the Middle Atlantic states and New England on 21 January behind a weak upper-level trough crossing the northeastern United States and eastern Canada. This pool of cold air was located beneath a region of confluence upwind of the 500-mb trough through 0000 UTC 22 January. Unlike many of the other cases described in this chapter, however, the cold air advection was associated with moderate westerly flow at 850 mb, not with strong northwesterly flow or with an intense surface anticyclone.

- The 0°C isotherm extended from the Texas coast across North Carolina on 21 January, prior to the development of the 850-mb low in the southeastern

# 850 MB

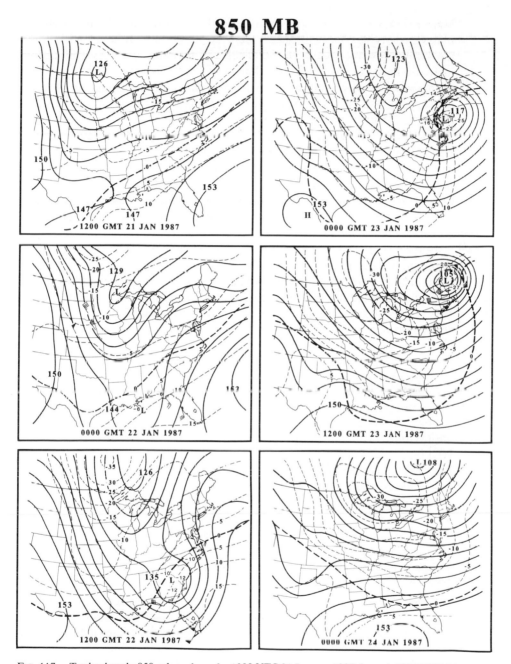

FIG. 117.　Twelve-hourly 850-mb analyses for 1200 UTC 21 January 1987 through 0000 UTC 24 January 1987. See Fig. 32 for details.

United States. As the cyclone moved northeastward, the 0°C isotherm remained nearly stationary through 1200 UTC 22 January. It then buckled into the familiar S-shaped pattern by 0000 UTC 23 January, as warm air advection increased in advance of the 850-mb low center and cold air advection increased to its rear. The increasing warm air advection from 1200 UTC 22 January through 0000 UTC 23 January was accompanied by heavy snowfall rates from the Carolinas to New England.

- The 850-mb low center deepened at an accelerating rate as it progressed northeastward from the Gulf of Mexico to Maine from 21 through 23 January. The deepening rate increased to $-90$ m $(12 \text{ h})^{-1}$ in the period ending at 1200 UTC 22 January, and to $-180$ m $(12 \text{ h})^{-1}$ in the following 12-hour period, as the cyclone moved northeastward along the Atlantic coast. The increased deepening rate was reflected in the amplifying local height-falls observed around the low center.

- A southerly LLJ developed along the Gulf coast at 0000 UTC 22 January and shifted to the Southeast coast by 1200 UTC 22 January, immediately upwind of the heaviest precipitation. As the storm intensified rapidly, easterly winds to the north of the 850-mb low center increased to 20–30 m s$^{-1}$ by 0000 and 1200 UTC 23 January. This jet was also directed toward the region where heavy precipitation was falling.

*500-mb geopotential height analyses*

- The cyclone developed as a deepening large-scale 500-mb trough was becoming established over the central and eastern United States on 21–22 January. A high-amplitude ridge was located over western North America. A 500-mb anticyclone became established over the North Atlantic Ocean east of Greenland on 21–22 January.

- Prior to cyclogenesis along the East Coast on 22 January, a region of confluence was observed behind a 500-mb trough moving rapidly along the northeastern United States–Canada border. Rising sea-level pressures and a weak surface anticyclone moved northeastward within this confluence zone.

- The surface cyclone developed in the Gulf of Mexico as a trough initially over Mexico began to move eastward and merged with the larger-scale trough developing over the central United States. Both troughs were marked by distinct 500-mb vorticity maxima, one located over the north-central United States and the other over northern Mexico at 1200 UTC 21 January. The southwestern trough lost its identity somewhat in the height and vorticity fields as it became embedded in or absorbed by the larger-wavelength trough deepening over central and eastern North America during 22 January. This relative "weakening" took place as rapid sea-level cyclonic development and heavy snowfall were occurring. Although the vorticity maximum diminished slightly in magnitude during this period, strong cyclonic vorticity advection was maintained downwind of the trough axis. By 1200 UTC 23 January, a distinct 500-mb height minimum and associated cyclonic vorticity maximum emerged north of New York, as the surface cyclone reached its lowest central pressure.

# 500 MB HEIGHT AND UPPER-LEVEL WINDS

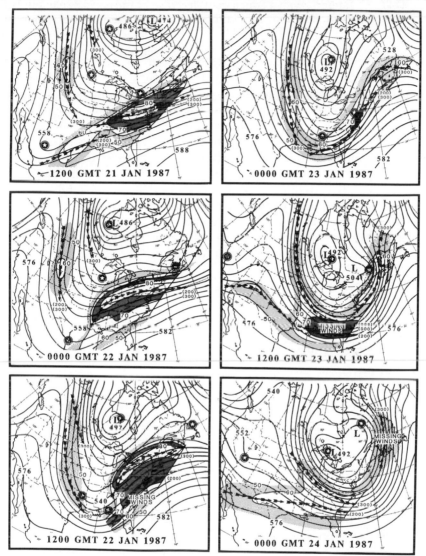

FIG. 118.   Twelve-hourly analyses of 500-mb geopotential heights and upper-level wind fields for 1200 UTC 21 January 1987 through 0000 UTC 24 January 1987. See Fig. 33 for details.

- A diffluent geopotential height pattern was not evident downwind of the trough axis until 0000 UTC 23 January, after the onset of rapid sea-level development.
- During the 24-hour period between 1200 UTC 22 January and 1200 UTC 23 January, the 500-mb trough rotated from a north–south line (from Wisconsin to Louisiana), to a northwest–southeast axis (a negative tilt), then to a west–east orientation (from southern Quebec to Maine). Concurrently, the surface low moved from Florida to Maine and deepened nearly 30 mb.

- The amplitude of the trough and downstream ridge increased through 0000 UTC 23 January, with a concurrent decrease of the half-wavelength between the axes.

*Upper-level wind analyses*

- A jet streak with wind speeds exceeding 80 m s$^{-1}$ was located at the base of the trough moving rapidly along the United States–Canada border on 21 January. As this trough continued eastward across southeastern Canada, wind speeds above 80 m s$^{-1}$ appeared to extend further southwestward toward the Gulf states. The surface anticyclone built northeastward from the south-central United States within the confluent entrance region of this intense jet streak.
- Two separate jet streaks were observed over the eastern United States by 1200 UTC 22 January, as rapid cyclogenesis commenced along the Southeast coast. The rapidly expanding area of heavy snowfall from Georgia to Maryland appeared to lie between the entrance region of the jet over the Northeast and the exit region of the jet that extended from the eastern Gulf of Mexico to southeastern Virginia. Unfortunately, the detailed structure of the jet streak along the East Coast was difficult to define at this time, since many upper-level wind reports were missing.
- A jet streak with wind speeds of at least 50–60 m s$^{-1}$ was located to the west of the deepening large-scale trough at all times. A separate jet streak located over central Canada through 1200 UTC 22 January was associated with a southward-propagating vortex that maintained very cold conditions over the northeastern United States during the following week.

*Infrared satellite imagery sequence*

- The infrared satellite images show a cloud pattern similar to that exhibited by the storm of 19–20 January 1978, with the rapid expansion of high clouds well to the north of the developing surface cyclone. The rapid expansion of the cloud mass occurred within the entrance region of the intense jet streak over the Ohio Valley. The northern and western edges of this cloud mass were aligned with the axis of the 80–90 m s$^{-1}$ upper-level jet extending from the Ohio Valley to New England at 0000 and 1200 UTC 22 January.
- At 1200 UTC 22 January, much of the heavy precipitation was located from beneath the highest cloud tops southward to the edge of the highest cloud tops over Georgia. By 0000 UTC 23 January, however, the heavy precipitation was located primarily under the warmer cloud-top region over the Middle Atlantic states and New England. A distinct comma shape defined the cloud mass at that time.
- A clear region began to appear over the northeastern Gulf of Mexico at 1201 UTC 22 January, southwest of the developing surface low. This dry air wedge expanded and raced northward in the following 12 hours, probably in association with the upper-level jet.
- It is difficult to distinguish the different airstreams in the images shown, but at 1201 UTC 22 January, a cold conveyor belt extended from near the surface east of Virginia southwest to the Carolinas, then rose sharply northwestward toward the western edge of the high clouds over Tennessee and Kentucky,

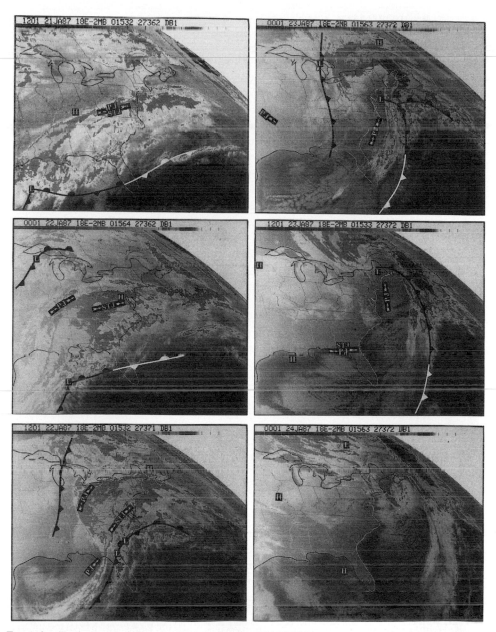

FIG. 119.  Twelve-hourly infrared satellite images for 1201 UTC 21 January 1987 through 0001 UTC 24 January 1987. Surface fronts and low- and high-pressure centers are also shown, along with the locations of upper-level polar (PJ) and subtropical (STJ) jet streaks.

and then turned north-northeastward parallel to the upper flow. Heavy precipitation was associated with this airstream, especially over the Carolinas and Virginia.

## 8.2    25–27 JANUARY 1987

*General remarks*

- The second major snowstorm in 4 days severely disrupted the Middle Atlantic region but missed most of New England, except southeastern Massachusetts. Although the heavy snow was not as widespread as in the previous storm, it combined with the deep snow cover already in place and some of the coldest temperatures of the season to produce severe winter weather conditions from Virginia to New Jersey. Washington, D.C., was at a standstill following these two storms, with schools closed for a week or more.
- Regions with snow accumulations exceeding 25 cm: southeastern West Virginia, central and northern Virginia, central and southern Maryland, Delaware (except the extreme north), southern New Jersey, eastern Long Island, and southeastern Massachusetts.
- Regions with snow accumulations exceeding 50 cm: none.
- Urban center snowfall amounts:

  | | |
  |---|---|
  | Washington, D.C.–Dulles Airport | 10.2″ (26 cm) |
  | Baltimore, MD | 9.6″ (24 cm) |
  | Philadelphia, PA | 4.1″ (10 cm) |
  | New York, NY–Avenue V, Brooklyn | 4.1″ (10 cm) |
  | Boston, MA | 3.3″ (8 cm) |

- Other selected snowfall amounts:

  | | |
  |---|---|
  | Patuxent River, MD | 18.0″ (46 cm) |
  | Salisbury, MD | 17.1″ (43 cm) |
  | Atlantic City, NJ | 16.3″ (41 cm) |
  | Edgartown, MA | 16.0″ (41 cm) |
  | Roanoke, VA | 13.9″ (35 cm) |
  | Dover, DE | 12.0″ (30 cm) |
  | Bridgehampton, NY | 11.0″ (28 cm) |

*The surface chart*

- Cold air for this snowfall was established over the north-central and north-eastern United States following the storm of 21–23 January. A broad area of high pressure (1025–1030 mb) extended from the northern Plains to the Middle Atlantic states and New England on 25–26 January. One high-pressure cell emerged and drifted from the upper Midwest southward to the Gulf coast on 26–27 January.
- Cold air damming, marked by an inverted sea-level pressure ridge, developed over the Middle Atlantic states by 0000 UTC 26 January. Coastal frontogenesis also occurred by this time along the coast of the Carolinas.
- The surface low advanced from eastern Texas to eastern Georgia in the 24-hour period ending at 0000 UTC 26 January, averaging 12 m s$^{-1}$. A center "jump" occurred at 0000 UTC 26 January as a new low developed along the coastal front over eastern South Carolina. This low moved to a position near Cape Hatteras, North Carolina, by 0600 UTC, then east-northeastward over

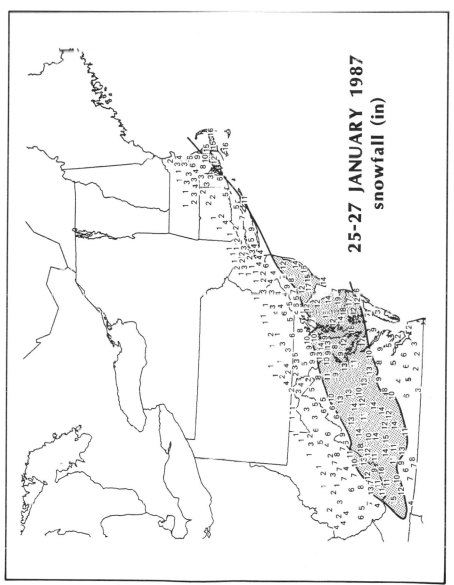

FIG. 120.   Snowfall (in) for 25–27 January 1987. See Fig. 3C for details.

# SURFACE

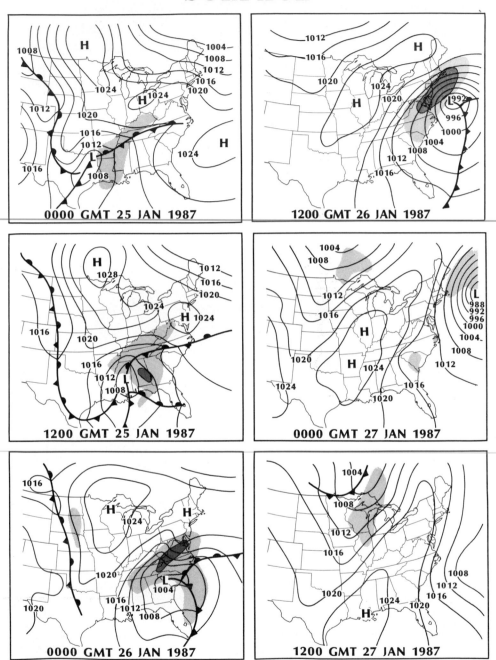

FIG. 121.    Twelve-hourly surface weather analyses for 0000 UTC 25 January 1987 through 1200 UTC 27 January 1987. See Fig. 31 for details.

the Atlantic Ocean, passing approximately 200 km southeast of southern New England. The center propagated at a rate of about 15 m s$^{-1}$ as it moved from South Carolina to a position about 340 km east of the Virginia coast in the 12-hour period ending at 1200 UTC 26 January.

- A band of light to moderate snowfall developed across the Middle Atlantic states on 25 January, north of a stationary front anchored over Tennessee and North Carolina. As the cyclone neared the coast, heavy snowfall developed from Virginia into southern Maryland, Delaware, and central and southern New Jersey. The heavy snow area expanded east-northeastward across eastern Long Island and extreme southeastern New England by 1200 UTC 26 January, as the storm moved out over the Atlantic Ocean.

- The cyclone deepened very slowly as it crossed the southeastern United States. The central sea-level pressure fell from 1006 to 1003 mb between 0000 UTC 25 January and 0000 UTC 26 January. As the surface low turned northeastward along the Carolina coast, a period of more significant deepening began. The central sea-level pressure fell from 1003 mb at 0000 UTC 26 January to 990 mb east of Virginia at 1200 UTC 26 January. This period of relatively rapid deepening coincided with the heavy snowfall across Virginia, Maryland, and Delaware.

- The heaviest snow fell within a well-defined 150-km-wide band centered parallel to and about 300 km to the north of the surface low track.

*The 850-mb chart*

- West to northwesterly flow and strong cold air advection brought an extensive area of $-10°C$ to $-25°C$ 850-mb temperatures over the Middle Atlantic states and New England by 25 January. A large temperature gradient had developed north of the 0°C isotherm, from the lower Ohio Valley across the Middle Atlantic states, by 0000 UTC 25 January.

- The 0°C isotherm generally extended from the 850-mb low center in the south-central United States eastward to the Carolina coast at 0000 and 1200 UTC 25 January. An S-shaped isotherm pattern became more evident by 0000 UTC 26 January, as the warm air advection east of the 850-mb low and cold air advection to the west both strengthened. The increasing warm air advection was accompanied by an increase of the temperature gradient along the Southeast coast, south of the 0°C isotherm. The intensifying temperature gradient and warm air advection were associated with the development of heavy snowfall over the Middle Atlantic states by 0000 UTC 26 January.

- The 850-mb low developed slowly, similar to its surface counterpart. Very little deepening was observed through 0000 UTC 26 January, with moderate deepening of $-60$ to $90$ m $(12 \text{ h})^{-1}$ thereafter.

- Commensurate with the slow deepening rates, the winds surrounding the 850-mb low were not as strong as those in many other cases. A 20–25 m s$^{-1}$ southerly LLJ was observed over the Gulf coast region through 1200 UTC 25 January, directed toward the expanding precipitation area east of the surface low. The largest 850-mb wind speeds observed in the heavy snow band north of the low center, however, did not exceed 20 m s$^{-1}$.

# 850 MB

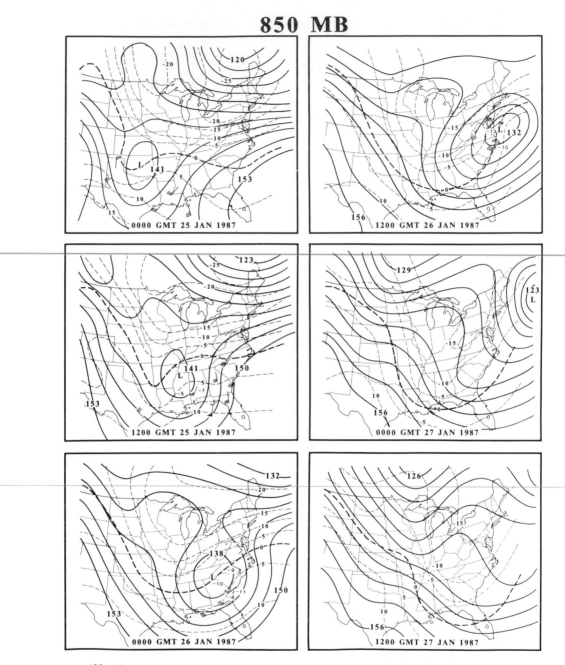

FIG. 122.    Twelve-hourly 850-mb analyses for 0000 UTC 25 January 1987 through 1200 UTC 27 January 1987. See Fig. 32 for details.

*500-mb geopotential height analyses*

- Prior to East Coast cyclogenesis, the 500-mb circulation was dominated by an intense vortex over eastern Canada which was associated with the East Coast storm a few days earlier and the cold air outbreak that followed it. A nearly stationary 500-mb anticyclone just east of Greenland slowed the movement of this vortex, maintaining cold air in the north-central and northeastern United States. The west coast of North America was characterized by a deamplifying ridge across Canada and diffluent flow entering the west coast. The trough that spawned this storm crossed the southwestern and south-central United States between 23 and 25 January in a westerly flow regime.

- The geopotential height field showed pronounced confluence across the northeastern United States through 26 January. This confluence zone was located above the sea-level high-pressure area and the west-northwesterly flow and cold air advection at 850 mb. The presence of a large 500-mb vortex over eastern Canada, a 500-mb anticyclone near Greenland, confluence across the northeastern United States, and a trough moving eastward through the southern United States produced a pattern similar to other snowstorms examined in chapter 7.

- The trough that was associated with the East Coast storm was marked by pronounced cyclonic curvature at the base as it propagated from Texas to the southeastern United States on 25 January. A distinct cyclonic vorticity maximum and its associated cyclonic vorticity advection pattern were observed with this trough. By 1200 UTC 26 January, however, the trough became more broadly curved and the vorticity pattern became more diffuse as the system moved eastward over the Atlantic Ocean. The 500-mb trough remained an open wave, with little increase in amplitude during cyclogenesis.

- The amplitude between the trough and its downstream ridge increased between 1200 UTC 25 January and 1200 UTC 26 January, a period marked by small, but increasing, surface deepening rates.

- Diffluence was evident downwind of the trough axis at all times. A northwest to southeast trough orientation (negative tilt) evolved on 25 January and was most prominent at 0000 UTC 26 January. The surface low and precipitation developed within the diffluent flow downwind of the trough axis.

*Upper-level wind analyses*

- A polar jet was located within the highly confluent flow over the northeastern United States on 25–26 January. Maximum wind speeds within this jet streak increased from 55 m s$^{-1}$ at 1200 UTC 25 January to 65 m s$^{-1}$ at 0000 UTC 26 January, and to greater than 70 m s$^{-1}$ by 1200 UTC 26 January. A band of snowfall developed across the Middle Atlantic states within the anticyclonic entrance region of the jet as it intensified.

- Separate jet features were evident upstream and downstream of the trough axis in the south-central United States during most of the observing period. A 60 m s$^{-1}$ wind maximum remained at the base of the trough as it propagated

# 500 MB HEIGHTS AND UPPER-LEVEL WINDS

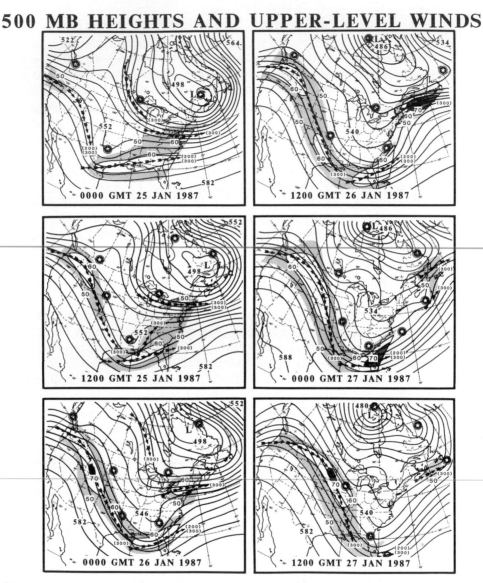

FIG. 123. Twelve-hourly analyses of 500-mb geopotential heights and upper-level wind fields for 0000 UTC 25 January 1987 through 1200 UTC 27 January 1987. See Fig. 33 for details.

toward the East Coast. At 1200 UTC 25 January and 0000 UTC 26 January, this jet streak exhibited a well-defined diffluent exit region over the southeastern United States, above the surface cyclogenesis. This jet/trough system in combination with the well-defined jet streak entrance region over the northeastern United States indicated the presence of the coupled transverse circulations that Uccellini and Kocin (1987) have linked to heavy snow events along the East Coast.

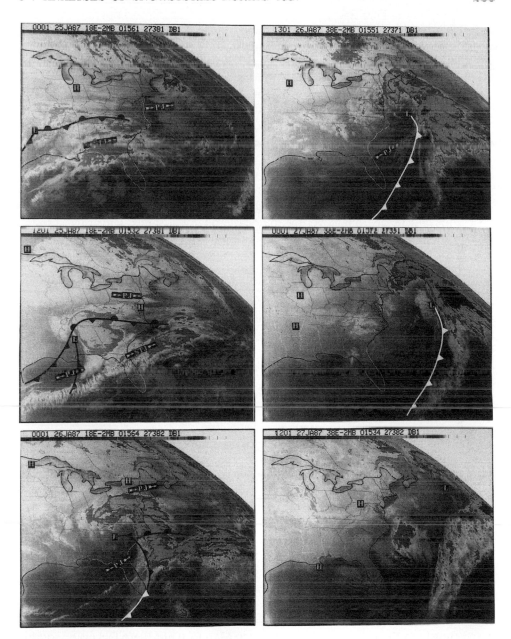

FIG. 124. Twelve-hourly infrared satellite images for 0001 UTC 25 January 1987 through 1201 UTC 27 January 1987. Surface fronts and low- and high-pressure centers are also shown, along with the locations of upper-level polar (PJ) and subtropical (STJ) jet streaks.

*Infrared satellite imagery sequence*

- By 1201 UTC 25 January, a cloud mass with cold cloud-top temperatures developed across Kentucky, Tennessee, Alabama, and Georgia, to the north and east of the surface low. These clouds were associated with an expanding area of light to moderate rain. A band of high cloud tops, though not as cold as the main cloud mass, streamed eastward from Kentucky across Virginia and New Jersey. This cloud area was associated with a band of light snow to the north of a stationary surface front and within the confluent entrance region of the upper-level jet streak extending from Pennsylvania across New England.

- A separate region of high clouds streaming eastward from the southeastern states at 0001 UTC 25 January moved well offshore in advance of the main comma cloud by 0001 UTC 26 January, in association with the subtropical jet.

- By 0001 UTC 26 January, the cloud mass from the Tennessee Valley had expanded northeastward and attained a comma shape. This cloud mass was located within the diffluent exit region of the upper-level jet streak rounding the base of the trough over the southeastern United States. The heaviest snow was falling over Virginia, near the southern edge of the coldest cloud tops that formed the comma head. The comma cloud continued to drift northeastward without any marked change in its shape through 1201 UTC 27 January.

- A clear tongue was observed in the southwestern quadrant of the developing cyclone after 0000 UTC 26 January, coincident with the polar jet located to the south of the surface low-pressure center.

## 8.3   22–24 FEBRUARY 1987

*General remarks*

- This rapidly developing cyclone produced an 8- to 12-hour period of very heavy, wet snowfall across portions of the Middle Atlantic states. The heavy loading of snow produced great damage to tree limbs and power lines, resulting in the loss of electricity to hundreds of thousands of people. The snowstorm occurred primarily during the night of 22–23 February, and came in stark contrast to the pleasant, mild, 10°C conditions of the preceding afternoon. An important aspect of this case is the widely varying operational numerical model forecasts of the location and intensity of cyclogenesis. Despite the inconsistency of the model forecasts and the unusually warm temperatures, some local forecasters made accurate predictions of the timing and amount of the snowfall.

- Regions with snow accumulations exceeding 25 cm: eastern West Virginia, northern Virginia, central and northern Maryland, northern Delaware, southeastern Pennsylvania, central and southern New Jersey, and part of Long Island.

- Regions with snow accumulations exceeding 50 cm: portions of southeastern Pennsylvania.

- Urban center snowfall amounts:
  Washington, D.C.–Dulles Airport                12.0″ (30 cm)
  Baltimore, MD                                          10.1″ (26 cm)

| | |
|---|---|
| Philadelphia, PA | 6.8″ (17 cm) |
| New York, NY–Kennedy Airport | 6.1″ (15 cm) |
| Boston, MA | — — |

- Other selected snowfall amounts:

| | |
|---|---|
| Coatesville, PA | 23.5″ (60 cm) |
| Clarksville, MD | 18.2″ (46 cm) |
| Wilmington, DE | 14.4″ (37 cm) |
| Martinsburg, WV | 13.3″ (34 cm) |
| Dover, DE | 10.5″ (27 cm) |

## The surface chart

- This snow event was preceded and followed by near- to slightly above-normal temperatures. Prior to the heavy snow, daytime temperatures on 22 February were generally between 8°C and 10°C, hardly typical for a major snowstorm. As the snow began, temperatures quickly fell to near freezing, but very few locations recorded readings lower than 0°C or −1°C. At Washington, D.C.–National Airport, 25 cm fell in spite of above-freezing air temperatures before and during the storm. The warm temperatures, snowfall rates of up to 13 cm h$^{-1}$, and unsteady wind speeds at the height of the storm enabled the snow to accumulate on a host of objects unable to withstand its weight. Tree limbs and power lines suffered extensive damage.

- The relatively mild temperatures preceding the storm were associated with a moderate anticyclone (1020–1025 mb) over the northeastern United States that weakened as the cyclone developed late on 22 February. Although the surface air mass was atypical for a heavy snow event, the high-pressure ridge extending southward east of the Appalachian Mountains at 1200 UTC 22 February and 0000 UTC 23 February suggested that cold air damming occurred prior to cyclogenesis. In addition, the upper-level temperature, wind, and geopotential height patterns were relatively typical of a heavy snowstorm.

- The cyclone which produced the heavy snow evolved from a complex system of low-pressure centers that consolidated into one rapidly developing storm along the Southeast coast. At 0000 UTC 22 February, a low-pressure system was developing in the western Gulf of Mexico while another low-pressure center and cold front were located over the northern Plains. In the following 24 hours, the cold front from the Plains states drifted to the Midwest as the low center over the Gulf of Mexico moved east-northeastward to Georgia, accompanied by an expanding shield of light to moderate rain.

- At 0000 UTC 23 February, rapid cyclogenesis and heavy snowfall were about to commence over the Middle Atlantic states. The Gulf low was crossing southern Georgia while two new lows formed. One center appeared over eastern Tennessee, immediately ahead of the upper-level vorticity maximum (see upper-level charts), as another center emerged off Charleston, South Carolina, along a developing coastal front. In the following 3 to 6 hours, the South Carolina low became the dominant cyclone.

- The low intensified rapidly after 0000 UTC 23 February, as it tracked northeastward along the coastal front across eastern North Carolina to southeastern

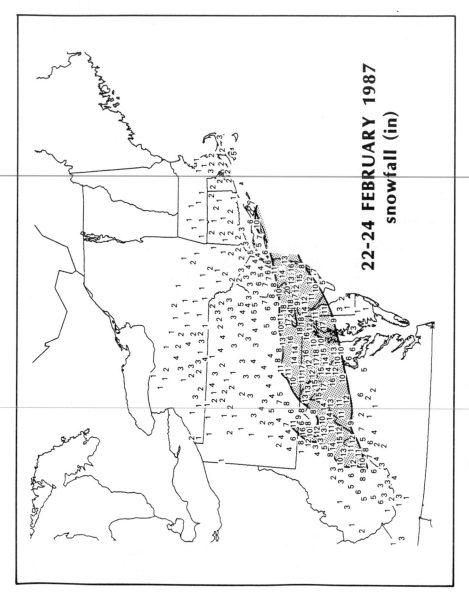

Fɪɢ. 125.    Snowfall (in) for 22–24 February 1987. See Fig. 30 for details.

# SURFACE

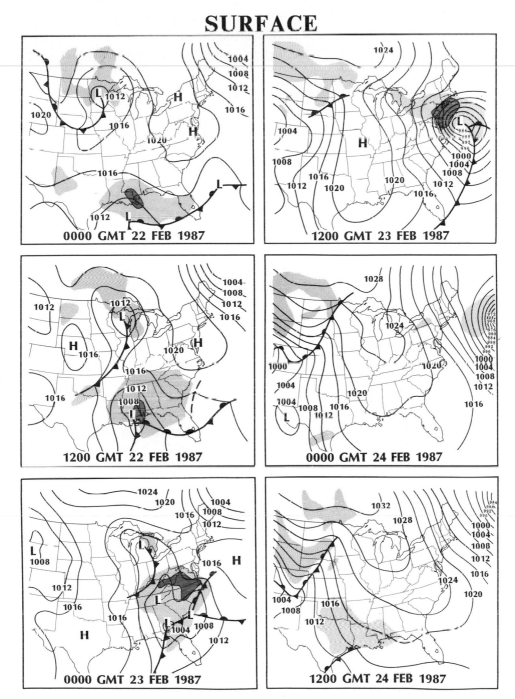

FIG. 126.    Twelve-hourly surface weather analyses for 0000 UTC 22 February 1987 through 1200 UTC 24 February 1987. See Fig. 31 for details.

Virginia. The central pressure fell from 1004 mb at 0000 UTC to 990 mb by 0900 UTC. The cyclone then followed a more easterly path off the Maryland coast and deepened at even greater rates, reaching 981 mb by 1200 UTC and 964 mb by 1800 UTC 23 February. The heavy snowfall from West Virginia and Virginia through Pennsylvania and New Jersey occurred between 0000 and 1800 UTC, during this period of rapid intensification. Lightning and thunder were also observed at many locations in the Middle Atlantic states early on 23 February.

- Winds increased along the Middle Atlantic coast as the cyclone deepened rapidly early on 23 February, but the wind speeds were not excessive, probably due to the relatively small preexisting sea-level pressure gradients and the isallobaric effects associated with the onset of rapid development.

- The snow melted quickly, as temperatures rose to 5°–10°C on the afternoon of 23 February.

*The 850-mb chart*

- A weak northwesterly flow regime and weak cold air advection at 850 mb characterized the precyclogenetic period over the northeastern United States on 21–22 February. This pattern was located beneath a region of upper-level confluent flow. The 850-mb temperatures north of the Virginia–North Carolina border ranged between 0°C and −10°C prior to the onset of the storm.

- Two 850-mb low centers were observed at 0000 UTC 22 February, one over Minnesota and another over Texas. The southern low center, which was associated with the surface low over the Gulf of Mexico, deepened significantly as it propagated northeastward. The northern low did not show any change in intensity until it disappeared after 0000 UTC 23 February.

- As the southern low moved northeastward, a pronounced S-shaped isotherm pattern developed over the southeastern United States, as warm air advection increased east and southeast of the center. This development accompanied the more rapid deepening of the low beginning at 0000 UTC 23 February and the outbreak of heavy snowfall in Virginia, West Virginia, and Maryland.

- Vigorous intensification of the 850-mb low commenced as the complex surface low-pressure area consolidated into one rapidly developing center along the Carolina coast. Deepening rates increased from −60 m (12 h)$^{-1}$ by 0000 UTC 23 February to −180 m (12 h)$^{-1}$ by 1200 UTC 23 February. The accelerating deepening rates were reflected in the local height changes, which amplified from −80 m (12 h)$^{-1}$ over the southern Appalachians at 0000 UTC 23 February to −210 m (12 h)$^{-1}$ over eastern Virginia by 1200 UTC 23 February.

- A southerly LLJ first appeared over Louisiana and Mississippi at 1200 UTC 22 February, as rain overspread the Gulf states. Wind speeds began to increase significantly over the southeastern United States by 0000 UTC 23 February, at the beginning of the cyclone's rapid deepening stage. An easterly wind component was observed across North Carolina and Virginia by 0000 UTC 23 February, directed toward the expanding area of heavy snow spreading across the Middle Atlantic states.

# 850 MB

FIG. 127.    Twelve-hourly 850-mb analyses for 0000 UTC 22 February 1987 through 1200 UTC 24 February 1987. See Fig. 32 for details.

*500-mb geopotential height analyses*

- The precyclogenetic environment at 500 mb was dominated by relatively stagnant features over eastern Canada, while conditions over the western and southern United States were changing rapidly. A large vortex was located near Newfoundland, where it had remained nearly stationary for the previous week. A 500-mb anticyclone was also relatively stationary from Hudson Bay to Greenland. These upper-level features were associated with a persistent north westerly flow at 850 mb and a sea-level high-pressure ridge across the northeastern United States and eastern Canada. This pattern maintained dry and relatively cold or cool surface air across the northeastern United States for more than a week.

- Farther west, the rapidly changing 500-mb height field featured a trough moving eastward from the southwestern United States, a separate trough propagating southward along the Pacific coast, and another trough drifting eastward across the northern United States. Each of these troughs was marked by a separate cyclonic vorticity maximum.

- The East Coast storm developed when the trough over the southwestern United States, which had been relatively stationary prior to 22 February, accelerated eastward, as the trough along the Pacific coast moved southward through California. The trough coming out of the southwestern United States initiated surface low development over the Gulf of Mexico, and later along the South Carolina coast.

- The trough that spawned the cyclone passed to the south and then east of a slower-moving trough over the northern United States on 22–23 February. These troughs phased by 0000 UTC 23 February and then developed a northwest-to-southeast orientation (a negative tilt). The rapidly deepening East Coast storm was located beneath the diffluent height contours east of the trough axis by 1200 UTC 23 February. The heavy precipitation and cyclogenesis occurred in advance of a well-defined cyclonic vorticity center, in the region of strong cyclonic vorticity advection.

- During the week preceding the storm, upper-level troughs progressing eastward across the United States encountered the large vortical circulation centered over eastern Canada and were suppressed and weakened. Operational numerical weather prediction models initialized prior to 1200 UTC 22 February forecast a continuation of this trend, as the trough over the upper Midwest was predicted to move at a faster rate than the southwestern trough. The larger phase speed forecast for the northern trough resulted in the complete deamplification of the southwestern trough as it reached the East Coast, with the subsequent cyclogenesis predicted to be very weak and suppressed to the south of its actual location. It therefore appears that the interactions of the two trough systems and their associated jet streaks were crucial to the evolution of this storm.

- As the trough from the southwestern United States moved eastward, the amplitude between the *upstream* ridge and trough decreased sharply by 0000 UTC 23 February. The amplitude of the trough and *downstream* ridge, however, remained the same or decreased only slightly. Although the amplitude of the trough and downstream ridge did not increase during rapid cyclogenesis

# 500 MB HEIGHTS AND UPPER-LEVEL WINDS

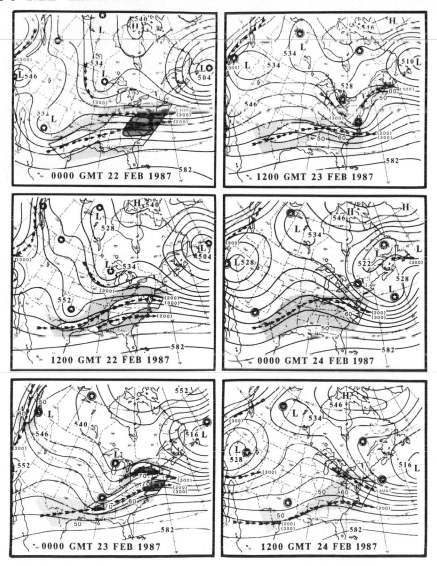

FIG. 128. Twelve-hourly analyses of 500-mb geopotential heights and upper-level wind fields for 0000 UTC 22 February 1987 through 1200 UTC 24 February 1987. See Fig. 33 for details.

between 0000 and 1200 UTC 23 February, the half-wavelength between the trough and downstream ridge decreased dramatically.

*Upper-level wind analyses*

- A subtropical jet streak with wind speeds in excess of 70 m s$^{-1}$ was located over the Middle Atlantic states prior to the storm.

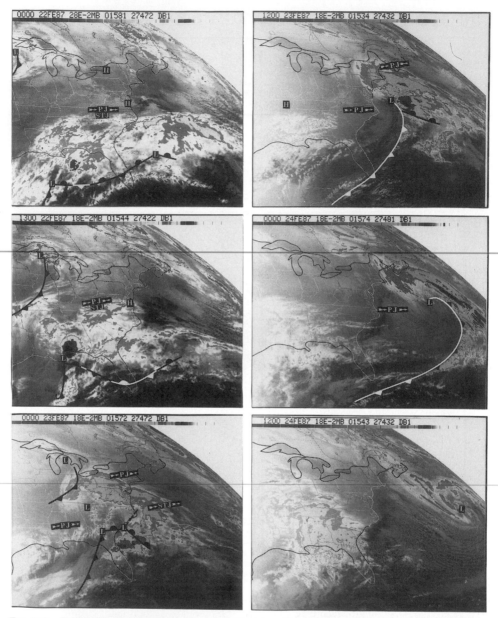

FIG. 129.   Twelve-hourly infrared satellite images for 0000 UTC 22 February 1987 through 1200 UTC 24 February 1987. Surface fronts and low- and high-pressure centers are also shown, along with the locations of upper-level polar (PJ) and subtropical (STJ) jet streaks.

- As the two troughs over the central United States approached the East Coast at 0000 UTC 23 February, wind speeds increased to greater than 70 m s$^{-1}$ within two separate jets, one over the eastern Great Lakes and New England, and the other extending from Texas across the Middle Atlantic states. This increase in wind speeds occurred as cyclogenesis was commencing along the South Carolina coast.

- The appearance of one or more jet streaks over the northeastern United States and a separate 70 m s$^{-1}$ jet maximum at the base of the trough over the Mississippi Valley at 0000 UTC 23 February produced the dual transverse circulations associated with the jet streaks, as described in section 5.3. The transverse circulations were well-defined in this system upstream of and prior to the major cyclone development, as shown by Uccellini and Kocin (1987).

*Infrared satellite imagery sequence*

- The satellite images initially showed a broad expanse of clouds across the southeastern United States, to the south of the upper-level jet system centered over the Middle Atlantic states. By 1300 UTC 22 February, patches of enhanced cold cloud tops were observed from Alabama to Missouri. The black-enhanced region near the Gulf coast depicts an outbreak of convection near the surface low center. The expanding cloud region from Mississippi to southern Missouri was associated with the upper-level trough propagating eastward across Texas.

- By 0000 UTC 23 February, a comma-shaped cloud mass had developed over the eastern United States. The arc-shaped comma head was located within the exit region of the jet streak at the base of the trough over Mississippi and Alabama. The comma head resembled the cold conveyor belt described in section 5.5, but there was no well-defined cloud signature to identify the warm conveyor belt.

- Rapid cyclogenesis was occurring within relatively shallow clouds over South Carolina at 0000 UTC 23 February. Although the highest cloud tops were located well to the north, an area of convective cloud tops was observed just south of the surface low center. Heavy snow developed from the Virginias to southern Pennsylvania, along the southern edge of the highest cloud tops associated with the comma head. By 1200 UTC 23 February, the comma-shaped cloud had become better organized, with the heaviest snow located near the southwestern edge of the comma head over New Jersey and Long Island. In the following 12 hours, the comma cloud became even more pronounced as the storm continued to deepen explosively over the ocean.

# References

Achtor, T. H., and L. H. Horn, 1986: Spring season Colorado cyclones. Part I: Use of composites to relate upper and lower tropospheric wind fields. *J. Climate Appl. Meteor.,* **25,** 732–743.

Anthes, R. A., and D. Keyser, 1979: Tests of a fine-mesh model over Europe and the United States. *Mon. Wea. Rev.,* **107,** 963–984.

——, Y.-H. Kuo and J. R. Gyakum, 1983: Numerical simulations of a case of explosive marine cyclogenesis. *Mon. Wea. Rev.,* **111,** 1174–1188.

Atlas, R., 1987: The role of oceanic fluxes and initial data in the numerical prediction of an intense coastal storm. *Dyn. Atmos. Oceans,* **10,** 359–388.

Austin, J. M., 1941: Favorable conditions for cyclogenesis near the Atlantic coast. *Bull. Amer. Meteor. Soc.,* **22,** 270.

——, 1951: Mechanisms of pressure change. *Compendium of Meteorology,* T. F. Malone, Ed., Amer. Meteor. Soc., 630–638.

Bailey, R. E., 1960: Forecasting of heavy snowstorms associated with major cyclones. J. J. George's *Weather Forecasting for Aeronautics,* Academic Press, 468–475.

Baker, D. G., 1970: A study of high pressure ridges to the east of the Appalachian Mountains. Ph.D. dissertation, Massachusetts Institute of Technology, 127 pp.

Ballentine, R. J., 1980: A numerical investigation of New England coastal frontogenesis. *Mon. Wea. Rev.,* **108,** 1479–1497.

Barnes, S. L., 1964: A technique for maximizing details in numerical weather map analysis. *J. Appl. Meteor.,* **3,** 396–409.

Beckman, S. K., 1987: Use of enhanced IR/visible satellite imagery to determine heavy snow areas. *Mon. Wea. Rev.,* **115,** 2060–2087.

Bell, G. D., and L. F. Bosart, 1988: Appalachian cold-air damming. *Mon. Wea. Rev.,* **116,** 137–161.

Bjerknes, J., 1919: On the structure of moving cyclones. *Geofys. Publ., Norske Videnskaps-Akad. Oslo 1,* **1,** 1–8.

——, 1951: Extratropical cyclones. *Compendium of Meteorology,* T. F. Malone, Ed., Amer. Meteor. Soc., 577–598.

——, 1954: The diffluent upper trough. *Arch. Meteor. Geophys. Bioklim.,* **A7,** 41–46.

——, and H. Solberg, 1922: Life cycle of cyclones and the polar front theory of atmospheric circulation. *Geofys. Publ., Norske Videnskaps-Akad. Oslo 3,* No. 1, 1–18.

——, and J. Holmboe, 1944: On the theory of cyclones. *J. Meteor.,* **1,** 1–22.

Bleck, R., 1977: Numerical simulation of lee cyclogenesis in the Gulf of Genoa. *Mon. Wea. Rev.,* **105,** 428–445.

Bonner, W. D., 1965: Statistical and kinematical properties of the low-level jet stream. Res. Paper, 38, Satellite and Mesometeorology Research Project, University of Chicago, 54 pp.

Bosart, L. F., 1975: New England coastal frontogenesis. *Quart. J. Roy. Meteor. Soc.,* **101,** 957–978.

——, 1981: The Presidents' Day snowstorm of 18–19 February 1979: A subsynoptic-scale event. *Mon. Wea. Rev.,* **109,** 1542–1566.

——, and S. C. Lin, 1984: A diagnostic analysis of the Presidents' Day storm of February 1979. *Mon. Wea. Rev.,* **112,** 2148–2177.

——, and F. Sanders, 1986: Mesoscale structure in the megalopolitan snowstorm of 11–12 February 1983, Part III: A large amplitude gravity wave. *J. Atmos. Sci.,* **43,** 924–939.

——, C. J. Vaudo and J. H. Helsdon, Jr., 1972: Coastal frontogenesis. *J. Appl. Meteor.,* **11,** 1236–1258.

Boucher, R. J., and R. J. Newcomb, 1962: Synoptic interpretation of some Tiros vortex patterns: A preliminary cyclone model. *J. Appl. Meteor.,* **1,** 127–136.

Boyle, J. S., and L. F. Bosart, 1986: Cyclone–anticyclone couplets over North America. Part II: Analysis of a major cyclone event over the eastern United States. *Mon. Wea. Rev.,* **114,** 2432–2465.

Brandes, E. A., and J. Spar, 1971: A search for necessary conditions for heavy snow on the East Coast. *J. Appl. Meteor.*, **11**, 397–409.

Brown, H. E., and D. A. Olson, 1978: Performance of NMC in forecasting a record breaking winter storm, 6–7 February 1978. *Bull. Amer. Meteor. Soc.*, **59**, 562–575.

Browne, R. F., and R. J. Younkin, 1970: Some relationships between 850 millibar lows and heavy snow occurrences over the central and eastern United States. *Mon. Wea. Rev.*, **98**, 399–401.

Browning, K. A., 1971: Radar measurements of air motion near fronts. *Weather*, **26**, 320–340.

——, 1986: Conceptual models of precipitation systems. *Wea. Forecasting*, **1**, 23–41.

——, and T. W. Harrold, 1969: Air motion and precipitation growth in a wave depression. *Quart. J. Roy. Meteor. Soc.*, **95**, 288–309.

Buzzi, A., and S. Tibaldi, 1978: Cyclogenesis in the lee of the Alps: A case study. *Quart. J. Roy. Meteor. Soc.*, **104**, 271–287.

Byers, H., 1965: *Elements of Cloud Physics.* University of Chicago Press, 191 pp.

Cahir, J. J., 1971: Implications of circulations in the vicinity of jet streaks at subsynoptic scales. Ph.D. thesis, Pennsylvania State University, 170 pp.

Caplovich, J., 1987: *Blizzard! The Great Storm of '88.* Vero Publishing Co., 242 pp.

Carlson, T. N., 1961: Lee-side frontogenesis in the Rocky Mountains. *Mon. Wea. Rev.*, **89**, 163–172.

——, 1980: Airflow through midlatitude cyclones and the comma cloud pattern. *Mon. Wea. Rev.*, **108**, 1498–1509.

Chang, C. B., D. J. Perkey and C. W. Kreitzberg, 1982: A numerical case study of the effects of latent heating on a developing wave cyclone. *J. Atmos. Sci.*, **39**, 1555–1570.

Charney, J. G., 1947: The dynamics of long waves in a baroclinic westerly current. *J. Meteor.*, **4**, 135–162.

Colucci, S. J., 1976: Winter cyclone frequencies over the eastern United States and adjacent western Atlantic, 1963–1973. *Bull. Amer. Meteor. Soc.*, **57**, 548–553.

——, 1985: Explosive cyclogenesis and large-scale circulation changes: Implications for atmospheric blocking. *J. Atmos. Sci.*, **42**, 2701–2717.

Danard, M. B., 1964: On the influence of released latent heat on cyclone development. *J. Appl. Meteor.*, **3**, 27–37.

——, and G. E. Ellenton, 1980: Physical influences on East Coast cyclogenesis. *Atmos.-Ocean.*, **18**, 65–82.

Danielsen, E. F., 1966: Research in four-dimensional diagnosis of cyclonic storm cloud systems. Rep. No. 66-30, Air Force Cambridge Res. Lab., Bedford, MA, 53 pp. [NTIS-AD-632668.]

Dines, W. H., 1925: The correlation between pressure and temperature in the upper air with a suggested explanation. *Quart. J. Roy. Meteor. Soc.*, **51**, 31–38.

Dirks, R. A., J. P. Kuettner and J. A. Moore, 1988: Genesis of Atlantic Lows Experiment (GALE): An overview. *Bull. Amer. Meteor. Soc.*, **69**, 148–160.

Eady, E. T., 1949: Long waves and cyclone waves. *Tellus*, **1**, 33–52.

Egger, J., 1974: Numerical experiments on lee cyclogenesis. *Mon. Wea. Rev.*, **102**, 847–860.

Eliassen, A., and E. Kleinschmidt, 1957: Dynamic meteorology. *Handbuch der Physik,* Vol. 48, S. Flugge, Ed., Springer-Verlag, 1–154.

Emanuel, K. A., 1985: Frontal circulations in the presence of small moist symmetric stability. *J. Atmos. Sci.*, **42**, 1062–1071.

Farrell, B., 1984: Modal and nonmodal baroclinic waves. *J. Atmos. Sci.*, **41**, 668–673.

——, 1985: Transient growth of damped baroclinic waves. *J. Atmos. Sci.*, **42**, 2718–2727.

Forbes, G. S., R. A. Anthes and D. W. Thompson, 1987: Synoptic and mesoscale aspects of an Appalachian ice storm associated with cold-air damming. *Mon. Wea. Rev.*, **115**, 564–591.

Foster, J. L., and R. J. Leffler, 1979: The extreme weather of February 1979 in the Baltimore–Washington area. *Nat. Wea. Dig.*, **4**, 16–21.

Gall, R., 1976: Structural changes of growing baroclinic waves. *J. Atmos. Sci.*, **33**, 374–390.

——, and D. R. Johnson, 1971: The generation of available potential energy by sensible heating: A case study. *Tellus*, **21**, 465–482.

Godev, N., 1971a: The cyclogenetic properties of the Pacific coast: Possible source of errors in numerical prediction. *J. Atmos. Sci.*, **28**, 968–972.

——, 1971b: Anticyclonic activity over south Europe and its relation to orography. *J. Appl. Meteor.*, **10**, 1097–1102.

Goree, P. A., and R. J. Younkin, 1966: Synoptic climatology of heavy snowfall over the central and eastern United States. *J. Appl. Meteor.*, **94**, 663–668.

Green, J. S. A., F. H. Ludlam and J. F. R. McIlveen, 1966: Isentropic relative-flow analysis and the parcel theory. *Quart. J. Roy. Meteor. Soc.*, **92**, 210–219.

Gyakum, J. R., 1983a: On the evolution of the QE II storm. Part I: Synoptic aspects. *Mon. Wea. Rev.*, **111**, 1137–1155.

——, 1983b: On the evolution of the QE II storm. Part II: Dynamic and thermodynamic structure. *Mon. Wea. Rev.*, **111**, 1156–1173.

Hadlock, R., and C. W. Kreitzberg, 1988: The Experiment on Rapidly Intensifying Cyclones over the Atlantic (ERICA) field study: Objectives and plans. *Bull. Amer. Meteor. Soc.*, **69**, 1309–1320.

Harrold, T. W., 1973: Mechanisms influencing the distribution of precipitation within baroclinic disturbances. *Quart. J. Roy. Meteor. Soc.*, **99**, 232–251.

Hayden, B. P., 1981: Secular variations in Atlantic coast extratropical cyclones. *Mon. Wea. Rev.*, **109**, 159–167.

Holton, J. R., 1979: *An Introduction to Dynamic Meteorology,* 2d edition, Academic Press, 391 pp.

Hoskins, B. J., M. E. McIntyre and A. W. Robertson, 1985: On the use and significance of isentropic potential vorticity maps. *Quart. J. Roy. Meteor. Soc.*, **111**, 877–946.

Hovanec, R. D., and L. H. Horn, 1975: Static stability and the 300 mb isotach field in the Colorado cyclogenetic area. *Mon. Wea. Rev.*, **103**, 628–638.

Johnson, D. R., and W. K. Downey, 1976: The absolute angular momentum budget of an extratropical cyclone: Quasi-Lagrangian diagnostics 3. *Mon. Wea. Rev.*, **104**, 3–14.

Kaplan, M. L., J. W. Zack, V. C. Wong and J. J. Tuccillo, 1982: A sixth-order mesoscale atmospheric simulation system applicable to research and real-time forecasting problems. *Proc., CIMMS 1982 Symposium,* Norman, OK, 38–84.

Keyser, D., and M. A. Shapiro, 1986: A review of the structure and dynamics of upper-level frontal zones. *Mon. Wea. Rev.*, **114**, 452–499.

——, and L. W. Uccellini, 1987: Regional models: Emerging research tools for synoptic meteorologists. *Bull. Amer. Meteor. Soc.*, **68**, 306–320.

Klein, W. H., and J. S. Winston, 1958: Geographical frequency of troughs and ridges on mean 700 mb charts. *Mon. Wea. Rev.*, **86**, 344–358.

Koch, S. E., M. L. desJardins and P. J. Kocin, 1983: An interactive Barnes objective map analysis scheme for use with satellite and conventional data. *J. Climate Appl. Meteor.*, **22**, 1487–1503.

Kocin, P. J., 1983: An analysis of the "Blizzard of '88." *Bull. Amer. Meteor. Soc.*, **64**, 1258–1272.

——, L. W. Uccellini, J. W. Zack and M. L. Kaplan, 1985: A mesoscale numerical forecast of an intense convective snowburst along the East Coast. *Bull. Amer. Meteor. Soc.*, **66**, 1412–1424.

——, —— and R. A. Petersen, 1986: Rapid evolution of a jet streak circulation in a pre-convective environment. *Meteor. Atmos. Phys.*, **35**, 103–138.

——, A. D. Weiss and J. J. Wagner, 1988: The great Arctic outbreak and East Coast blizzard of February 1899. *Wea. Forecasting*, **3**, 305–318.

Krishnamurti, T. N., 1968: A study of a developing wave cyclone. *Mon. Wea. Rev.*, **96**, 208–217.

Lai, C.-C., and L. F. Bosart, 1988: A case study of trough merger in split westerly flow. *Mon. Wea. Rev.*, **116**, 1838–1856.

Leese, J. A., 1962: The role of advection in the formation of vortex cloud patterns. *Tellus*, **14**, 409–421.

Ludlum, D. M., 1966: *Early American Winters I 1604–1820.* American Meteorological Society, 283 pp.

——, 1968: *Early American Winters II 1821–1870.* American Meteorological Society, 257 pp.

——, 1976: *The Country Journal New England Weather Book.* Houghton Mifflin Company, 147 pp.

——, 1982: *The American Weather Book.* Houghton Mifflin Company, 296 pp.

——, 1983: *The New Jersey Weather Book.* Rutgers University Press, 249 pp.

MacDonald, B. C., and E. R. Reiter, 1988: Explosive cyclogenesis over the eastern United States. *Mon. Wea. Rev.*, **116**, 1568–1586.

Marks, F. D., and P. M. Austin, 1979: Effects of the New England coastal front on the distribution of precipitation. *Mon. Wea. Rev.*, **107**, 53–67.

Mather, J. R., H. Adams and G. A. Yoshioka, 1964: Coastal storms of the eastern United States. *J. Appl. Meteor.*, **3**, 693–706.

Mattocks, C., and R. Bleck, 1986: Jet streak dynamics and geostrophic adjustment processes during the initial stages of lee cyclogenesis. *Mon. Wea. Rev.*, **114**, 2033–2056.

Miller, J. E., 1946: Cyclogenesis in the Atlantic coastal region of the United States. *J. Meteor.*, **3**, 31–44.

Mook, C. P., and K. S. Norquest, 1956: The heavy snowstorm of 18–19 March 1956. *Mon. Wea. Rev.*, **84**, 116–125.

Mullen, S. L., 1983: Explosive cyclogenesis associated with cyclones in polar air streams. *Mon. Wea. Rev.*, **111**, 1537–1553.

Murray, R., and S. M. Daniels, 1953: Transverse flow at entrance and exit to jet streams. *Quart. J. Roy. Meteor. Soc.*, **79**, 236–241.

Namias, J., and P. F. Clapp, 1949: Confluence theory of the high tropospheric jet stream. *J. Meteor.*, **6**, 330–336.

Newton, C. W., 1956: Mechanisms of circulation change during a lee cyclogenesis. *J. Meteor.*, **13**, 528–539.

——, and A. V. Persson, 1962: Structural characteristics of the subtropical jet stream and certain lower-stratospheric wind systems. *Tellus*, **14**, 221–241.

——, and A. Trevisan, 1984: Clinogenesis and frontogenesis in jet stream waves. Part II: Channel model numerical experiments. *J. Atmos. Sci.*, **41**, 2735–2755.

Orlanski, I., 1975: A rational subdivision of scales for atmospheric processes. *Bull. Amer. Meteor. Soc.*, **56**, 527–530.

Palmén, E., 1951: The aerology of extratropical disturbances. *Compendium of Meteorology*, T. Malone, Ed., Amer. Meteor. Soc., 599–620.

——, and C. W. Newton, 1969: *Atmospheric Circulation Systems*. Academic Press, 603 pp.

Petterssen, S., 1955: A general survey of factors influencing development at sea-level. *J. Meteor.*, **12**, 36–42.

——, 1956: *Weather Analysis and Forecasting*, Vol. 1. McGraw-Hill, 428 pp.

——, D. L. Bradbury and K. Pedersen, 1962: The Norwegian cyclone models in relation to heat and cold sources. *Geofys. Publ. Norske Viderskaps-Akad. Oslo*, **24**, 243–280.

Reitan, C. H., 1974: Frequencies of cyclones and cyclogenesis for North America, 1951–1970. *Mon. Wea. Rev.*, **102**, 861–868.

Reiter, E. R., 1963. *Jet Stream Meteorology* The University of Chicago Press, 515 pp.

——, 1969: Tropospheric circulations and jet streams. *World Survey of Climatology*, Vol. 4, Climate of the Free Atmosphere, D. F. Rex, Ed., 85–203.

Richwien, B. A., 1980: The damming effect of the southern Appalachians. *Nat. Wea. Dig.*, **5**, 2–12.

Riehl, H., J. Badner, J. E. Hovda, N. E. LaSeur, L. L. Means, W. C. Palmer, M. J. Schroeder, L. W. Snellman and others, 1952: Forecasting in the middle latitudes. *Meteor. Monogr.*, No. 5, Amer. Meteor. Soc., 80 pp.

Roebber, P. J., 1984: Statistical analysis and updated climatology of explosive cyclones. *Mon. Wea. Rev.*, **112**, 1577–1589.

Rogers, E., and L. F. Bosart, 1986: An investigation of explosively deepening oceanic cyclones. *Mon. Wea. Rev.*, **114**, 702–718.

Rosenblum, H. S., and F. Sanders, 1974: Meso-analysis of a coastal snowstorm in New England. *Mon. Wea. Rev.*, **102**, 433–442.

Sanders, F., 1986a: Explosive cyclogenesis in the west-central North Atlantic Ocean, 1981–84. Part I: Composite structure and mean behavior. *Mon. Wea. Rev.*, **114**, 1781–1794.

——, 1986b: Frontogenesis and symmetric stability in a major New England snowstorm. *Mon. Wea. Rev.*, **114**, 1847–1862.

——, and J. R. Gyakum, 1980: Synoptic dynamic climatology of the bomb. *Mon. Wea. Rev.*, **108**, 1589–1606.

——, and L. F. Bosart, 1985a: Mesoscale structure in the megalopolitan snowstorm of 11–12 February 1983. Part I: Frontogenetical forcing and symmetric instability. *J. Atmos. Sci.*, **42**, 1050–1061.

——, and ——, 1985b: Mesoscale structure in the megalopolitan snowstorm of 11–12 February 1983. Part II: Doppler radar study of the New England snowband. *J. Atmos. Sci.*, **42**, 1398–1407.

Sanderson, A. N., and R. B. Mason, Jr., 1958: Behavior of two East Coast storms, 13–24 March 1958. *Mon. Wea. Rev.*, **86**, 109–115.

Scherhag, R., 1937: Bermerkurgen über die bedeutung der konvergenzen und divergenzen du geschwindig-keitsfeldes fur die Druckänderungen. *Beitr. Physik Atmosphare*, **24**, 122–129.

Seltzer, M. A., R. E. Passarelli and K. A. Emanuel, 1985: The possible role of symmetric instability in the formation of precipitation bands. *J. Atmos. Sci., 42,* 2207–2219.

Shapiro, M. A., and P. J. Kennedy, 1981: Research aircraft measurements of jet stream geostrophic and ageostrophic winds. *J. Atmos. Sci., 38,* 2642–2652.

Sinclair, M. R., and R. L. Ellsberry, 1986: A diagnostic study of baroclinic disturbances in polar air streams. *Mon. Wea. Rev., 114,* 1957–1983.

Smith, R. B., 1979: The influence of the mountains on the atmosphere. *Advances in Geophysics, 21,* 87–230.

——, 1984: A theory of lee cyclogenesis. *J. Atmos. Sci., 41,* 1159–1168.

Spiegler, D. B., and G. E. Fisher, 1971: A snowfall prediction method for the Atlantic seaboard. *Mon. Wea. Rev., 99,* 311–325.

Staley, D. O., and R. L. Gall, 1977: On the wavelength of maximum baroclinic instability. *J. Atmos. Sci., 34,* 1679–1688.

Stauffer, D. R., and T. T. Warner, 1987: A numerical study of Appalachian cold-air damming and coastal frontogenesis. *Mon. Wea. Rev., 115,* 799–821.

Stewart, R. E., R. W. Shaw and G. A. Isaac, 1987: Canadian Atlantic Storms Program: The meteorological field project. *Bull. Amer. Meteor. Soc., 68,* 338–345.

Sutcliffe, R. C., 1939: Cyclonic and anticyclonic development. *Quart. J. Roy. Meteor. Soc., 65,* 518–524.

——, 1947: A contribution to the problem of development. *Quart. J. Roy. Meteor. Soc., 73,* 370–383.

——, and A. G. Forsdyke, 1950: The theory and use of upper air thickness patterns in forecasting. *Quart. J. Roy. Meteor. Soc., 76,* 189–217.

Tracton, M. S., 1973: The role of cumulus convection in the development of extratropical cyclones. *Mon. Wea. Rev., 101,* 573–593.

Uccellini, L. W., 1986: The possible influence of upstream upper-level baroclinic processes on the development of the QE II storm. *Mon. Wea. Rev., 114,* 1019–1027.

——, and D. R. Johnson, 1979: The coupling of upper- and lower-tropospheric jet streaks and implications for the development of severe convective storms. *Mon. Wea. Rev., 107,* 682–703.

——, and P. J. Kocin, 1987: An examination of vertical circulations associated with heavy snow events along the East Coast of the United States. *Wea. Forecasting, 2,* 289–308.

——, P. J. Kocin, R. A. Petersen, C. H. Wash and K. F. Brill, 1984: The Presidents' Day cyclone of 18–19 February 1979: Synoptic overview and analysis of the subtropical jet streak influencing the precyclogenetic period. *Mon. Wea. Rev., 112,* 31–55.

——, D. Keyser, K. F. Brill and C. H. Wash, 1985: The Presidents' Day cyclone of 18–19 February 1979: Influence of upstream trough amplification and associated tropopause folding on rapid cyclogenesis. *Mon. Wea. Rev., 113,* 962–988.

——, R. A. Petersen, K. F. Brill, P. J. Kocin and J. J. Tuccillo, 1987: Synergistic interactions between an upper-level jet streak and diabatic processes that influence the development of a low-level jet and a secondary coastal cyclone. *Mon. Wea. Rev., 115,* 2227–2261.

Wagner, A. J., 1957: Mean temperature from 1000 mb to 500 mb as a predictor of precipitation type. *Bull. Amer. Meteor. Soc., 38,* 584–590.

Weiss, A. D., and J. J. Wagner, 1987: *A Catalogue of Philadelphia Snowstorms since 1888.* Unpublished manuscript.

Werstein, I., 1960: *The Blizzard of '88.* Thomas Y. Crowell Co., 157 pp.

Whitaker, J. S., L. W. Uccellini and K. F. Brill, 1988: A model-based diagnostic study of the explosive development phase of the Presidents' Day cyclone. *Mon. Wea. Rev., 116,* 2337–2365.

Widger, W. K., Jr., 1964: A synthesis of interpretations of extratropical vortex patterns as seen by TIROS. *Mon. Wea. Rev., 92,* 263–282.

Wolfsberg, D. G., K. A. Emanuel and R. E. Passarelli, 1986: Band formation in a New England winter storm. *Mon. Wea. Rev., 114,* 1552–1569.

Young, M. V., G. A. Monk and K. A. Browning, 1987: Interpretation of satellite imagery of a rapidly deepening cyclone. *Quart. J. Roy. Meteor. Soc., 113,* 1089–1115.